ASE Test Preparation

Medium/Heavy Duty Truck Technician Certification Series

Suspension and Steering (T5)

5th Edition

Australia • Brazil • Japan • Korea • Mexico • Singapore • Spain • United Kingdom • United States

ASE Test Preparation: Medium/Heavy Duty Truck Technician Certification Series, Suspension and Steering (T5), 5th Edition

Vice President, Technology and Trades Professional Business Unit: Gregory L. Clayton

Director, Professional Transportation Industry Training Solutions: Kristen L. Davis

Product Manager: Katie McGuire

Editorial Assistant: Danielle Filippone

Director of Marketing: Beth A. Lutz

Senior Marketing Manager: Jennifer Barbic

Senior Production Director: Wendy Troeger

Production Manager: Sherondra Thedford

Senior Art Director: Benjamin Gleeksman

Content Project Management: PreMediaGlobal

Section Opener Image: © Kevin Norris/ Shutterstock.com

© 2013 Delmar, Cengage Learning

ALL RIGHTS RESERVED. No part of this work covered by the copyright herein may be reproduced, transmitted, stored or used in any form or by any means graphic, electronic, or mechanical, including but not limited to photocopying, recording, scanning, digitizing, taping, Web distribution, information networks, or information storage and retrieval systems, except as permitted under Section 107 or 108 of the 1976 United States Copyright Act, without the prior written permission of the publisher.

For product information and technology assistance, contact us at **Cengage Learning Customer & Sales Support, 1-800-354-9706**

For permission to use material from this text or product, submit all requests online at **www.cengage.com/permissions**

Further permissions questions can be emailed to **permissionrequest@cengage.com**

ISBN-13: 978-1-111-12901-9

ISBN-10: 1-111-12901-0

Delmar
Executive Woods
5 Maxwell Drive
Clifton Park, NY 12065
USA

Cengage Learning is a leading provider of customized learning solutions with office locations around the globe, including Singapore, the United Kingdom, Australia, Mexico, Brazil, and Japan. Locate your local office at **www.cengage.com/global**

Cengage Learning products are represented in Canada by Nelson Education, Ltd.

For more information on transportation titles available from Delmar, Cengage Learning, please visit our website at **www.trainingbay.cengage.com.**

For more learning solutions, please visit our corporate website at **www.cengage.com.**

Notice to the Reader
Publisher does not warrant or guarantee any of the products described herein or perform any independent analysis in connection with any of the product information contained herein. Publisher does not assume, and expressly disclaims, any obligation to obtain and include information other than that provided to it by the manufacturer. The reader is expressly warned to consider and adopt all safety precautions that might be indicated by the activities described herein and to avoid all potential hazards. By following the instructions contained herein, the reader willingly assumes all risks in connection with such instructions. The publisher makes no representations or warranties of any kind, including but not limited to, the warranties of fitness for particular purpose or merchantability, nor are any such representations implied with respect to the material set forth herein, and the publisher takes no responsibility with respect to such material. The publisher shall not be liable for any special, consequential, or exemplary damages resulting, in whole or part, from the readers' use of, or reliance upon, this material.

Printed in the United States of America
2 3 4 5 6 7 18 17 16 15 14

Table of Contents

Preface

Delmar, a part of Cengage Learning, is very pleased that you have chosen to use our ASE Test Preparation Guide to help prepare yourself for the Suspension and Steering (T5) ASE certification examination. This guide is designed to help prepare you for your actual exam by providing you with an overview and introduction of the testing process, introducing you to the task list for the Suspension and Steering (T5) certification exam, giving you an understanding of what knowledge and skills you are expected to have in order to successfully perform the duties associated with each task area, and providing you with several preparation exams designed to emulate the live exam content in hopes of assessing your overall exam readiness.

If you have a basic working knowledge of the discipline you are testing for, you will find this book is an excellent guide, helping you understand the "must know" items needed to successfully pass the ASE certification exam. This manual is not a textbook. Its objective is to prepare the individual who has the existing requisite experience and knowledge to attempt the challenge of the ASE certification process. This guide cannot replace the hands-on experience and theoretical knowledge required by ASE to master the vehicle repair technology associated with this exam. If you are unable to understand more than a few of the preparation questions and their corresponding explanations in this book, it could be that you require either more shop-floor experience or further study.

This book begins by providing an overview of, and introduction to, the testing process. This section outlines what we recommend you do to prepare, what to expect on the actual test day, and overall methodologies for your success. This section is followed by a detailed overview of the ASE task list to include explanations of the knowledge and skills you must possess to successfully answer questions related to each particular task. After the task list, we provide six sample preparation exams for you to use as a means of evaluating areas of understanding, as well as areas requiring improvement in order to successfully pass the ASE exam. Delmar is the first and only test preparation organization to provide so many unique preparation exams. We enhanced our guides to include this support as a means of providing you with the best preparation product available. Section 6 of this guide includes the answer keys for each preparation exam, along with the answer explanations for each question. Each answer explanation also contains a reference back to the related task or tasks that it assesses. This will provide you with a quick and easy method for referring back to the task list whenever needed. The last section of this book contains blank answer sheet forms you can use as you attempt each preparation exam, along with a glossary of terms.

OUR COMMITMENT TO EXCELLENCE

Thank you for choosing Delmar, Cengage Learning for your ASE test preparation needs. All of the writers, editors, and Delmar staff have worked very hard to make this test preparation guide second to none. We feel confident that you will find this guide easy to use and extremely beneficial as you prepare for your actual ASE exam.

Delmar, Cengage Learning has sought out the best subject-matter experts in the country to help with the development of *ASE Test Preparation: Medium/Heavy Duty Truck Technician Certification Series: Suspension and Steering (T5), 5th Edition*. Preparation questions are authored and then reviewed by a group of certified subject-matter experts to ensure the highest level of quality and validity to our product.

If you have any questions concerning this guide or any guide in this series, please visit us on the web at **http://www.trainingbay.cengage.com**.

For web-based online test preparation for ASE certifications, please visit us on the web at **http://www.techniciantestprep.com/**to learn more.

ABOUT THE AUTHOR

Michael Meredith has over 35 years experience in the heavy diesel repair industry. During his tenure, he has worked in fleet maintenance management; truck, trailer, and heavy equipment repair; and has experience is suspension, frame straightening and repair, and alignment. He is currently a member of the Fleet Maintenance Education team of FedEx Freight. In addition to his work experience, he has been an instructor of diesel technologies for Denver Automotive and Diesel Technical College and is an ASE Master certified medium/heavy truck technician.

ABOUT THE SERIES ADVISOR

Brian (BJ) Crowley has experienced several different aspects of the diesel industry over the past 10 years. Now a diesel technician in the oil and gas industry, BJ owned and operated a diesel repair shop where he repaired heavy, medium, and light trucks, in addition to agricultural and construction equipment. He earned an Associate's degree in diesel technology from Elizabethtown Community and Technical College and is an ASE Master certified medium/heavy truck technician.

SECTION 1

The History and Purpose of ASE

ASE began as the National Institute for Automotive Service Excellence (NIASE). It was founded as a nonprofit, independent entity in 1972 by a group of industry leaders with the single goal of providing a means for consumers to distinguish between incompetent and competent technicians. It accomplishes this goal through the testing and certification of repair and service professionals. Though it is still known as the National Institute for Automotive Service Excellence, it is now called "ASE" for short.

Today, ASE offers more than 40 certification exams in automotive, medium/heavy-duty truck, collision repair and refinish, school bus, transit bus, parts specialist, automobile service consultant, and other industry-related areas. At this time, there are more than 385,000 professionals nationwide with current ASE certifications. These professionals are employed by new car and truck dealerships, independent repair facilities, fleets, service stations, franchised service facilities, and more.

ASE's certification exams are industry-driven and cover practically every on-highway vehicle service segment. The exams are designed to stress the knowledge of job-related skills. Certification consists of passing at least one exam and documenting two years of relevant work experience. To maintain certification, those with ASE credentials must be re-tested every five years.

While ASE certifications are a targeted means of acknowledging the skills and abilities of an individual technician, ASE also has a program designed to provide recognition for highly qualified repair, support, and parts businesses. The Blue Seal of Excellence Recognition Program allows businesses to showcase their technicians and their commitment to excellence. One of the requirements of becoming Blue Seal recognized is that the facility must have a minimum of 75 percent of its technicians ASE certified. Additional criteria apply, and program details can be found on the ASE website.

ASE recognized that educational programs serving the service and repair industry also needed a way to be recognized as having the faculty, facilities, and equipment to provide a quality education to students wanting to become service professionals. Through the combined efforts of ASE, industry, and education leaders, the nonprofit organization entitled the National Automotive Technicians Education Foundation (NATEF) was created in 1983 to evaluate and recognize academic programs. Today more than 2,000 educational programs are NATEF certified.

For additional information about ASE, NATEF, or any of their programs, the following contact information can be used:

National Institute for Automotive Service Excellence (ASE)
101 Blue Seal Drive S.E.
Suite 101
Leesburg, VA 20175
Telephone: 703-669-6600
Fax: 703-669-6123
Website: **www.ase.com**

SECTION

2 Overview and Introduction

Participating in the National Institute for Automotive Service Excellence (ASE) voluntary certification program provides you with the opportunity to demonstrate you are a qualified and skilled professional technician who has the "know-how" required to successfully work on today's modern vehicles.

EXAM ADMINISTRATION

> *Note:* After November 2011, ASE will no longer offer paper and pencil certification exams. There will be no Winter testing window in 2012, and ASE will offer and support CBT testing exclusively starting in April 2012.

ASE provides computer-based testing (CBT) exams, which are administered at test centers across the nation. It is recommended that you go to the ASE website at ***http://www.ase.com*** and review the conditions and requirements for this type of exam. There is also an exam demonstration page that allows you to personally experience how this type of exam operates before you register.

CBT exams are available four times annually, for two-month windows, with a month of no testing in between each testing window:

- January/February – Winter testing window
- April/May – Spring testing window
- July/August – Summer testing window
- October/November – Fall testing window

Please note, testing windows and timing may change. It is recommended you go to the ASE website at ***http://www.ase.com*** and review the latest testing schedules.

UNDERSTANDING TEST QUESTION BASICS

ASE exam questions are written by service industry experts. Each question on an exam is created during an ASE-hosted "item-writing" workshop. During these workshops, expert service representatives from manufacturers (domestic and import), aftermarket parts and equipment manufacturers, working technicians, and technical educators gather to share ideas and convert them into actual exam questions. Each exam question written by these experts must then survive review by all members of the group. The questions are designed to address the practical application of repair and diagnosis knowledge and skills practiced by technicians in their day-to-day work.

After the item-writing workshop, all questions are pre-tested and quality-checked on a national sample of technicians. Those questions that meet ASE standards of quality and accuracy are included in the scored sections of the exams; the "rejects" are sent back to the drawing board or discarded altogether.

Depending on the topic of the certification exam, you will be asked between 40 and 80 multiple-choice questions. You can determine the approximate number of questions you can expect to be asked during the Suspension and Steering (T5) certification exam by reviewing the task list in Section 4 of this book. The five-year recertification exam will cover this same content; however, the number of questions for each content area of the recertification exam will be reduced by approximately one-half.

> ***Note:*** Exams may contain questions that are included for statistical research purposes only. Your answers to these questions will not affect your score, but since you do not know which ones they are, you should answer all questions in the exam.

Using multiple criteria, including cross sections by age, race, and other background information, ASE is able to guarantee that exam questions do not include bias for or against any particular group. A question that shows bias toward any particular group is discarded.

TEST-TAKING STRATEGIES

Before beginning your exam, quickly look over the exam to determine the total number of questions that you will need to answer. Having this knowledge will help you manage your time throughout the exam to ensure you have enough available to answer all of the questions presented. Read through each question completely before marking your answer. Answer the questions in the order they appear on the exam. Leave the questions blank that you are not sure of and move on to the next question. You can return to those unanswered questions after you have finished the others. These questions may actually be easier to answer at a later time, once your mind has had additional time to consider them on a subconscious level. In addition, you might find information in other questions that will help you recall the answers to some of them.

Multiple-choice exams are sometimes challenging because there are often several choices that may seem possible, or partially correct, and therefore it may be difficult to decide on the most appropriate answer choice. The best strategy, in this case, is to first determine the correct answer before looking at the answer options. If you see the answer you decided on, you should still be careful to examine the other answer options to make sure that none seems more correct than yours. If you do not know or are not sure of the answer, read each option very carefully and try to eliminate those options that you know are incorrect. That way, you can often arrive at the correct choice through a process of elimination.

If you have gone through the entire exam, and you still do not know the answer to some of the questions, ***then guess.*** Yes, guess. You then have at least a 25 percent chance of being correct. While your score is based on the number of questions answered correctly, any question left blank, or unanswered, is automatically scored as incorrect.

There is a lot of "folk" wisdom on the subject of test taking that you may hear about as you prepare for your ASE exam. For example, there are those who would advise you to avoid response options that use certain words such as ***all***, ***none***, ***always***, ***never***, ***must***, and ***only***, to name a few. This, they claim, is because nothing in life is exclusive. They would advise you to choose response options that use words that allow for some exception, such as ***sometimes***, ***frequently***, ***rarely***, ***often***, ***usually***, ***seldom***, and ***normally***. They would also advise you to avoid the first and last option (A or D) because exam writers, they feel, are more comfortable if they put the correct answer in the middle (B or C) of the choices. Another recommendation often offered is to select the option that is either shorter or longer than the other three choices because it is more likely to be correct. Some would advise you to never change an answer since your first intuition is usually correct. Another area of "folk" wisdom focuses specifically on any repetitive patterns created by your question responses (e.g., A, B, C, A, B, C, A, B, C).

Many individuals may say that there are actual grains of truth in this "folk" wisdom, and whereas with some exams, this may prove true, it is not relevant in regard to the ASE certification exams. ASE validates all exam questions and test forms through a national sample of technicians, and only those questions and test forms that meet ASE standards of quality and accuracy are included in the scored sections of the exams. Any biased questions or patterns are discarded altogether, and therefore, it is highly unlikely you will experience any of this "folk" wisdom on an actual ASE exam.

PREPARING FOR THE EXAM

Delmar, Cengage Learning wants to make sure we are providing you with the most thorough preparation guide possible. To demonstrate this, we have included hundreds of preparation questions in this guide. These questions are designed to provide as many opportunities as possible to prepare you to successfully pass your ASE exam. The preparation approach we recommend and outline in this book is designed to help you build confidence in demonstrating what task area content you already know well while also outlining what areas you should review in more detail prior to the actual exam.

We recommend that your first step in the preparation process should be to thoroughly review Section 3 of this book. This section contains a description and explanation of the type of questions you will find on an ASE exam.

Once you understand how the questions will be presented, we then recommend that you thoroughly review Section 4 of this book. This section contains information that will help you establish an understanding of what the exam will be evaluating, and specifically, how many questions to expect in each specific task area.

As your third preparatory step, we recommend you complete your first preparation exam, located in Section 5 of this book. Answer one question at a time. After you answer each question, review the answer and question explanation information located in Section 6. This section will provide you with instant response feedback, allowing you to gauge your progress, one question at a time, throughout this first preparation exam. If after reading the question explanation you do not feel you understand the reasoning for the correct answer, go back and review the task list overview (Section 4) for the task that is related to that question. Included with each question explanation is a clear identifier of the task area that is being assessed (e.g., Task A.1). If at that point you still do not feel you have a solid understanding of the material, identify a good source of information on the topic, such as an educational course, textbook, or other related source of topical learning, and do some additional studying.

After you have completed your first preparation exam and have reviewed your answers, you are ready to complete your next preparation exam. A total of six practice exams are available in Section 5 of this book. For your second preparation exam, we recommend that you answer the questions as if you were taking the actual exam. Do not use any reference material or allow any interruptions in order to get a feel for how you will do on the actual exam. Once you have answered all of the questions, grade your results using the Answer Key in Section 6. For every question that you gave an incorrect answer to, study the explanations to the answers and/or the overview of the related task areas. Try to determine the root cause for missing the question. The easiest thing to correct is learning the correct technical content. The hardest things to correct are behaviors that lead you to an incorrect conclusion. If you knew the information but still got the question incorrect, there is likely a test-taking behavior that will need to be corrected. An example of this would be reading too quickly and skipping over words that affect your reasoning. If you can identify what you did that caused you to answer the question incorrectly, you can eliminate that cause and improve your score.

Here are some basic guidelines to follow while preparing for the exam:

- Focus your studies on those areas you are weak in.
- Be honest with yourself when determining if you understand something.
- Study often but for short periods of time.
- Remove yourself from all distractions when studying.
- Keep in mind that the goal of studying is not just to pass the exam; the real goal is to learn.
- Prepare physically by getting a good night's rest before the exam, and eat meals that provide energy but do not cause discomfort.
- Arrive early to the exam site to avoid long waits as test candidates check in.
- Use all of the time available for your exams. If you finish early, spend the remaining time reviewing your answers.
- Do not leave any questions unanswered. If absolutely necessary, guess. All unanswered questions are automatically scored as incorrect.

Here are some items you will need to bring with you to the exam site:

- A valid government or school-issued photo ID
- Your test center admissions ticket
- A watch (not all test sites have clocks)

> ***Note:*** Books, calculators, and other reference materials are not allowed in the exam room. The exceptions to this list are English-Foreign dictionaries or glossaries. All items will be inspected before and after testing.

WHAT TO EXPECT DURING THE EXAM

When taking a CBT exam, as soon as you are seated in the testing center, you will be given a brief tutorial to acquaint you with the computer-delivered test prior to taking your certification exam(s). The CBT exams allow you to select only one answer per question. You can also change your answers as many times as you like. When you select a second answer choice, the CBT will automatically unselect your first answer choice. If you want to skip a question to return to later, you can utilize the "flag" feature, which will allow you to quickly identify and review questions whenever you are ready. Prior to completing your exam, you will also be provided with an opportunity to review your answers and address any unanswered questions.

TESTING TIME

Each individual ASE CBT exam has a fixed time limit. Individual exam times will vary based upon exam area and will range anywhere from a half hour to two hours. You will also be given an additional 30 minutes beyond what is allotted to complete your exams to ensure you have adequate time to perform all necessary check-in procedures, complete a brief CBT tutorial, and potentially complete a post-test survey.

You can register for and take multiple CBT exams during one testing appointment. The maximum time allotment for a CBT appointment is four and a half hours. If you happen to register for so many exams that you will require more time than this, your exams will be scheduled into multiple appointments. This could mean that you have testing on both the morning and afternoon of the same day, or they could be scheduled on different days, depending on your personal preference and the test center's schedule.

It is important to understand that if you arrive late for your CBT test appointment, you will not be able to make up any missed time. You will only have the scheduled amount of time remaining in your appointment to complete your exam(s).

Also, while most people finish their CBT exams within the time allowed, others might feel rushed or not be able to finish the test due to the implied stress of a specific, individual time limit allotment. Before you register for the CBT exams, you should review the number of exam questions that will be asked along with the amount of time allotted for that exam to determine whether you feel comfortable with the designated time limitation or not.

As an overall time management recommendation, you should monitor your progress and set a time limit you will follow with regard to how much time you will spend on each individual exam question. This should be based on the total number of questions you will be answering.

Also, it is very important to note that if for any reason you wish to leave the testing room during an exam, you must first ask permission. If you happen to finish your exam(s) early and wish to leave the testing site before your designated session appointment is completed, you are permitted to do so only during specified dismissal periods.

UNDERSTANDING HOW YOUR EXAM IS SCORED

You can gain a better perspective about the ASE certification exams if you understand how they are scored. ASE exams are scored by an independent organization having no vested interest in ASE or in the automotive industry. With CBT exams, you will receive your exam scores immediately.

Each question carries the same weight as any other question. For example, if there are 50 questions, each is worth 2 percent of the total score. The passing grade is 70 percent.

Your exam results can tell you:

- Where your knowledge equals or exceeds that needed for competent performance, or
- Where you might need more preparation.

Your ASE exam score report is divided into content "task" areas; it will show the number of questions in each content area and how many of your answers were correct. These numbers provide information about your performance in each area of the exam. However, because there may be a different number of questions in each content area of the exam, a high percentage of correct answers in an area with few questions may not offset a low percentage in an area with many questions.

It should be noted that one does not "fail" an ASE exam. The technician who does not pass is simply told "More Preparation Needed." Though large differences in percentages may indicate problem areas, it is important to consider how many questions were asked in each area. Since each exam evaluates all phases of the work involved in a service specialty, you should be prepared in each area. A low score in one area could keep you from passing an entire exam. If you do not pass the exam, you may take it again at any time it is scheduled to be administered.

There is no such thing as average. You cannot determine your overall exam score by adding the percentages given for each task area and dividing by the number of areas. It does not work that way because there generally are not the same number of questions in each task area. A task area with 20 questions, for example, counts more toward your total score than a task area with 10 questions.

Your exam report should give you a good picture of your results and a better understanding of your strengths and areas needing improvement for each task area.

SECTION

3 Types of Questions on an ASE Exam

Understanding not only what content areas will be assessed during your exam, but how you can expect exam questions to be presented will enable you to gain the confidence you need to successfully pass an ASE certification exam. The following examples will help you recognize the types of question styles used in ASE exams and assist you in avoiding common errors when answering them.

Most initial certification tests are made up of between 40 to 80 multiple-choice questions. The five-year recertification exams will cover the same content as the initial exam; however, the actual number of questions for each content area will be reduced by approximately one-half. Refer to Section 4 of this book for specific details regarding the number of questions to expect during the initial Suspension and Steering (T5) certification exam.

Multiple-choice questions are an efficient way to test knowledge. To correctly answer them you must consider each answer choice as a possibility, and then choose the answer choice that *best* addresses the question. To do this, read each word of the question carefully. Do not assume you know what the question is asking until you have finished reading the entire question.

About 10 percent of the questions on an actual ASE exam will reference an illustration. These drawings contain the information needed to correctly answer the question. The illustration should be studied carefully before attempting to answer the question. When the illustration is showing a system in detail, look over the system and try to figure out how the system works before you look at the question and the possible answers. This approach will ensure that you do not answer the question based upon false assumptions or partial data, but instead have reviewed the entire scenario being presented.

MULTIPLE-CHOICE/DIRECT QUESTIONS

The most common type of question used on an ASE exam is the direct multiple-choice style question. This type of question contains an introductory statement, called a stem, followed by four options: three incorrect answers, called distracters, and one correct answer, the key.

When the questions are written, the point is to make the distracters plausible to draw an inexperienced technician to inadvertently select one of them. This type of question gives a clear indication of the technician's knowledge.

Here is an example of a direct-style question:

TASK A.4

1. What size steering wheel is most commonly used on heavy-duty trucks?

   A. 18 inches

   B. 20 inches

   C. 22 inches

   D. 24 inches

Answer A is incorrect. Heavy-duty trucks/tractors do not use 18 inch steering wheels.

Answer B is incorrect. Original equipment manufacturers (OEMs) specify a 22 inch steering wheel for heavy-duty trucks.

Answer C is correct. The most common size steering wheel is 22 inches.

Answer D is incorrect. 24 inch steering wheels are not used on heavy-duty trucks.

COMPLETION QUESTIONS

A completion question is similar to the direct question except the statement may be completed by any one of the four options to form a complete sentence.

Here's an example of a completion question:

1. Continual excessive steering effort on a power steering system may be caused by:

   A. Rough worm shaft bearings.

   B. A loose worm shaft bearing preload.

   C. Low pump pressure.

   D. Air in the power steering system.

TASK A.5

Answer A is incorrect. Rough worm shaft bearings may exhibit noise during operation, but not excessive steering effort.

Answer B is incorrect. Loose worm shaft bearing preload would cause steering instability.

Answer C is correct. Low pump pressure would cause a continual excessive steering effort.

Answer D is incorrect. Air in the power steering system would not cause continual excessive steering effort. It will cause pump noise due to cavitation and erratic steering assist.

TECHNICIAN A, TECHNICIAN B QUESTIONS

This type of question is usually associated with an ASE exam. It is, in fact, two true-false statements grouped together, such as: "Technician A says..." and "Technician B says...," followed by "Who is correct?"

In this type of question, you must determine whether either, both, or neither of the statements is correct. To answer this type of question correctly, you must carefully read each technician's statement and judge it on its own merit.

Sometimes this type of question begins with a statement about some analysis or repair procedure. This statement provides the setup or background information required to understand the conditions about which Technician A and Technician B are talking, followed by two statements about the cause of the concern, proper inspection, identification, or repair choices.

Analyzing this type of question is a little easier than the other types because there are only two ideas to consider, although there are still four choices for an answer.

Again, Technician A, Technician B questions are really double true-or-false statements. The best way to analyze this type of question is to consider each technician's statement separately. Ask yourself, "Is A true or false? Is B true or false?" Once you have completed an individual evaluation of each statement, you will have successfully determined the correct answer choice for the question, "Who is correct?"

An important point to remember is that an ASE Technician A, Technician B question will never have Technician A and B directly disagreeing with each other. That is why you must evaluate each statement independently.

An example of a Technician A/Technician B-style question looks like this:

TASK B.11

1. U-bolt clamping force is being checked. Technician A states that when the U-bolt is hit with a brass hammer, if the sound being made is a dull thud, the U-bolt torque is OK. Technician B says that the U-bolt torque should be checked at regular intervals as specified by the OEM. Who is correct?

   A. A only
   B. B only
   C. Both A and B
   D. Neither A nor B

Answer A is incorrect. Technician B is also correct.

Answer B is incorrect. Technician A is also correct.

Answer C is correct. Both Technicians are correct. A dull thud sound indicates that the U-bolts are not loose. However, this should not be the only check made. Checking the torque at specified intervals ensures that the proper torque is maintained.

Answer D is incorrect. Both Technicians are correct.

EXCEPT QUESTIONS

Another type of question used on ASE exams contains answer choices that are all correct except for one. To help easily identify this type of question, whenever it is presented in an exam, the word "EXCEPT" will always be displayed in capital letters. Furthermore, a cautionary statement will alert you to the fact that the next question is different from the ones otherwise found in the exam. With the EXCEPT type of question, only one **incorrect** choice will actually be listed among the options, and that incorrect choice will be the key to the question. That is, the incorrect statement is counted as the correct answer for that question.

Be careful to read these question types slowly and thoroughly; otherwise you may overlook what the question is actually asking and answer the question by selecting the first correct statement.

An example of this type of question would appear as follows:

TASK A.5

1. A driver complains of excessive steering wheel freeplay. All of the following may be the cause EXCEPT:

   A. Worn wheel bearings.
   B. Worn drag link sockets.
   C. Worn tie rod ends.
   D. Worn steering shaft universal joints.

Answer A is correct. Worn wheel bearings would cause noise and roughness when manually rotating the hub. They would not have an effect on the steering wheel freeplay.

Answer B is incorrect. Worn drag link sockets would require the steering wheel to be rotated farther in before the linkage would move.

Answer C is incorrect. As with any ball-type socket, worn tie rod ends would allow for play in the steering linkage. It would result in excessive play in the steering wheel.

Answer D is incorrect. Worn steering shaft U-joints would allow for excess movement in the steering shaft and steering wheel.

LEAST LIKELY QUESTIONS

LEAST LIKELY questions are similar to EXCEPT questions. Look for the answer choice that would be the LEAST LIKELY cause (most incorrect) for the described situation. To help easily identify these types of questions, whenever they are presented in an exam the words "LEAST LIKELY" will always be displayed in capital letters. In addition, you will be alerted before a LEAST LIKELY question is posed. Read the entire question carefully before choosing your answer.

An example of this type of question is shown here:

1. The LEAST LIKELY cause for continual excessive steering effort on a power steering system is:

TASK A.5

   A. Rough worm shaft bearings.
   B. Improper worm shaft bearing preload.
   C. Low pump pressure.
   D. Air in the power steering system.

Answer A is correct. Rough worm shaft bearings may exhibit noise during operation, but not excessive steering effort.

Answer B is incorrect. If worm bearing preload is adjusted too tight, the result would be a binding of the steering gear causing excessive steering effort.

Answer C is incorrect. Low pump pressure would cause a continual excessive steering effort.

Answer D is incorrect. Air in the power steering system would not cause continual excessive steering effort. It will cause pump noise due to cavitation and erratic steering assist.

SUMMARY

The question styles outlined in this section are the only ones you will encounter on any ASE certification exam. ASE does not use any other types of question styles, such as fill-in-the-blank, true/false, word-matching, or essay. ASE also will not require you to draw diagrams or sketches to support any of your answer selections, although any of the described question styles may include illustrations, charts, or schematics to clarify a question. If a formula or chart is required to answer a question, it will be provided for you.

SECTION

4 Task List Overview

INTRODUCTION

This section of the book outlines the content areas or *task list* for this specific certification exam, along with a written overview of the content covered in the exam.

The task list describes the actual knowledge and skills necessary for a technician to successfully perform the work associated with each skill area. This task list is the fundamental guideline you should use to understand what areas you can expect to be tested on, as well as how each individual area is weighted to include the approximate number of questions you can expect to be given for that area during the ASE certification exam. It is important to note that the number of exam questions for a particular area is to be used as a guideline only. ASE advises that the questions on the exam may not equal the number specifically listed on the task list. The task lists are specifically designed to tell you what ASE expects you to know how to do and to help prepare you to be tested.

Similar to the role this task list will play in regard to the actual ASE exam, Delmar, Cengage Learning has developed six preparation exams, located in Section 5 of this book, using this task list as a guide. It is important to note that although both ASE and Delmar, Cengage Learning use the same task list as a guideline for creating these test questions, none of the test questions you will see in this book will be found in the actual, live ASE exams. This is true for any test preparatory material you use. Real exam questions are *only* visible during the actual ASE exams.

Task List at a Glance

The Suspension and Steering (T5) task list focuses on four core areas, and you can expect to be asked a total of approximately 50 questions on your certification exam, broken out as outlined:

A. Steering System Diagnosis and Repair (12 questions)
B. Suspension, Frame, and 5th Wheel Diagnosis and Repair (16 questions)
C. Wheel Alignment Diagnosis, Adjustment, and Repair (13 questions)
D. Wheels, Tires, and Hub Diagnosis and Repair (9 questions)

Based upon this information, the following is a general guideline demonstrating which areas will have the most focus on the actual certification exam. This data may help you prioritize your time when preparing for the exam.

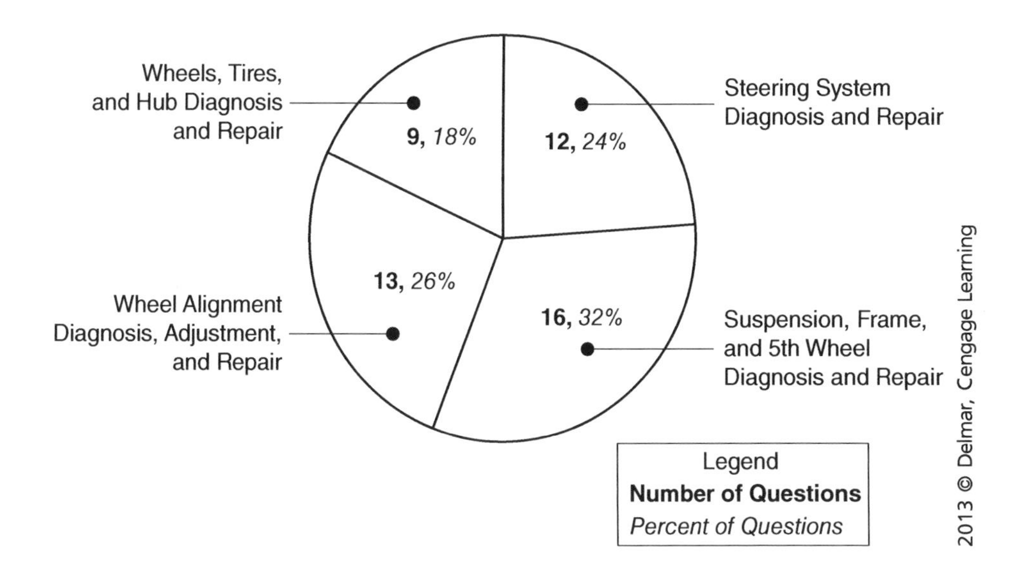

> *Note:* The actual number of questions you will be given on the ASE certification exam may vary slightly from the information provided in the task list, as exams may contain questions that are included for statistical research purposes only. Don't forget that your answers to these research questions will not affect your score.

TASK LIST AND OVERVIEW

A. Steering System Diagnosis and Repair (12 Questions)

1. Diagnose steering column (tilt, telescoping, or fixed) for noise, looseness, and binding problems; determine needed repairs.

Steering column component malfunctions often contribute to binding, hard steering, looseness, and excessive steering wheel play. Establish the complaint by getting as much information as possible from the driver. Next, give the vehicle a thorough road test.

When experiencing excessive wander, check for steering column shaft adjustment, worn upper or lower bearings, or worn U-joint or flex coupling.

Worn steering column bushings or bearings leave excessive lash in the linkage that can cause a loose feeling in the steering wheel when driving straight, noise when hitting a bump in the road because the bushing acts as a cushion, and a pull to the left or right while driving because the wheels will drift to the pitch of the highway.

When experiencing no steering recovery, look for lack of lubrication in the U-joint or flex coupling, worn upper or lower bearings, bent steering shaft, or U-joints not installed properly.

When experiencing excessive steering wheel movement, the cause might be a loose steering wheel, worn U-joint or flex coupling, loose steering gear, a loose pitman arm, loose or worn steering linkage components, or steering gear out of adjustment.

When experiencing steering column binding, inspect for worn or seized U-joint or flex couplings, steering column bearings binding or misaligned, bent steering shaft, or U-joints not properly installed.

If component repair or replacement is necessary, refer to OEM procedures.

2. Inspect and replace steering shaft U-joint(s), slip joints, bearings, bushings, and seals; phase shaft U-joints.

Most steering systems use universal joints (U-joints) or flex couplings to allow the shafts to mount on an angle. The intermediate shaft, used on most truck steering systems, runs from the lower end of the steering column shaft to the steering gear input shaft. The intermediate shaft connects to the steering gear input shaft with either a flex coupling or U-joint. In some cases, the upper end of the intermediate shaft is also equipped with a U-joint. With this configuration, the lower coupler assembly can be rebuilt or is replaceable while the upper coupler assembly is not.

To prevent the possibility of speed fluctuations caused by shaft U-joints operating at an angle, proper phasing is required. Most steering shafts require phasing both shaft yokes on the same plane during installation. This ensures that the steering gear and the steering wheel operate in unison even though the shaft speed may fluctuate. Always check the manufacturer's service manuals to confirm the proper phasing procedure.

When testing a linkage joint for excessive wear, you should not compress the steering column U-joint with a C-clamp or large pliers. This process will compress the tension spring in the joint and give the impression there is wear in the joint. Simply push against the joint with the force you can create with your hands. This should be enough to identify excessive wear.

Steering column bearings and bushings should be checked for wear, noisy operation, and binding during operation. In addition, steering shaft seals and boots should be inspected for proper installation and damage.

Since steering shaft, joint, and bearing service varies depending on the steering column type, always refer to the instructions in the manufacturer's service manual for the proper repair/replacement procedure.

3. Check cab mounting and adjust ride height.

Cab air suspensions contain one or two air springs, two shock absorbers, and a leveling valve. The leveling valve maintains the proper air pressure in the air springs to provide the correct cab height. You adjust the height control or leveling valve (similar to air suspension systems) to maintain the proper cab height. Note: Suspension height should be set before any diagnostics are performed.

A visual inspection of the air bags should include looking for cracking or wear of the rubber component, piston condition and mounting, dirt accumulation around the piston, and exterior components that may come in contact with the bag or interfere with bag travel. The shock absorbers should be checked for proper mounting, leakage, and proper travel. In addition, the leveling valve should be inspected for proper mounting, leakage, and proper ride height adjustment. Always check the manufacturer's service manuals to confirm the proper ride height.

4. Remove the steering wheel (including steering wheels equipped with electrical/electronic controls and components); install and center the steering wheel. Inspect, test, replace, and calibrate steering angle sensors.

To remove the steering wheel, begin by removing the hub cover and steering wheel retaining nut. Using a suitable puller, remove the steering wheel from the upper steering column shaft. Reinstall the steering wheel nut to prevent the steering column shaft from sliding out of the jacket tube (if equipped).

Caution: Do not use a knock-off type steering wheel puller or strike the end of the upper shaft with a hammer. Damage to the steering shaft bearing may occur.

When the front wheels are in the straight-ahead position, the steering wheel should be centered. If the steering wheel is not centered, the sector shaft may not be contacting the high point on the ball nut. This condition causes erratic steering control. If the steering wheel is not centered, be sure the steering wheel is installed in the proper position on the steering shaft. After a toe adjustment, the steering wheel may require centering.

On an idler arm-type steering linkage, the steering wheel is centered by rotating the tie rod sleeves with the proper tie rod sleeve rotating tool. If the steering wheel spoke is low on the left side, rotate the tie rod sleeves to shorten the left tie rod and lengthen the right tie rod. A one-quarter turn on the tie rod sleeves moves the steering wheel approximately 1 inch.

On a heavy-duty truck steering system, centering the steering wheel may be accomplished through an adjustable drag link if equipped. If not equipped with an adjustable drag link, ensure the front wheels are in a straight-ahead position. Disconnect the drag link from the pitman arm. Center the steering gear and check to see if the pitman arm is splined in the proper position. Check the measurement between the pitman arm bore and the steering control arm. Check the length of the fixed drag link to ensure it is the proper component. If correct, reinstall the drag link and then reposition the steering wheel on the steering shaft.

Note: If a caster adjustment is required, it should be done before centering the steering system and wheel.

Because advanced ABS ECUs are individually customized based on expected sensor locations and orientations, an out-of-calibration sensor or incorrectly positioned sensor may lead to unwanted and/or unneeded stability interventions, which can result in incidents leading to loss of vehicle control.

The steering angle sensor must be recalibrated as part of:

- Steering wheel replacement (see Caution at the end of this list)
- Steering angle sensor replacement
- Any maintenance that involves opening the connector hub from the steering angle sensor to the column
- Any maintenance or repair work on the steering linkage, steering gear, or other related mechanism
- Wheel alignment or wheel track adjustment
- Accident repairs where damage to the steering angle sensor or assembly, or any part of the steering system may have occurred

Caution: When replacing a steering wheel, use only OEM-approved steering wheels. Take care not to damage the steering angle sensor or interfere with its operation during installation.

5. Diagnose power steering system noise, steering binding, darting/oversteer, reduced wheel cut, steering wheel kick, pulling, non-recovery, turning effort, looseness, hard steering, overheating, fluid leakage, and fluid aeration problems; determine needed repairs.

Power steering system noise, steering binding, darting/oversteer, reduced wheel cut, steering wheel kick, pulling, non-recovery, turning effort, looseness, hard steering, overheating, fluid leakage, and fluid aeration problems may be caused by a variety of mechanical and/or hydraulic issues. In addition, load factors and road conditions may also have an effect on the steering components. Before inspecting the components, gather as much information as you can. If possible, interview the driver to determine if the issue is ongoing or only occurs under certain conditions. Road test the vehicle under the conditions described and try to duplicate the condition. Pay particular attention to speed, load factors, directional issues, and noisy operation. Tire inflation and suspension component condition may also play a role in steering system issues.

Noise: Can be caused by various sources. A visual and audible inspection must first be performed to determine cause (check belts, fluid level, mounts, etc.).

Binding and Turning Effort, Hard Steering: Can be the result of improper torque applied to any newly installed components; seized or underlubricated moving components; improper phasing during the installation of shafts and/or U-joints; loose drive belts; misalignment due to wear, fatigue, or collision damage; or improper servicing of power steering filters.

Looseness: Improper torque specifications (i.e., steering box preload) and worn or improperly installed components can lead to excessive steering wheel movement.

Darting/oversteer: Loose steering components and/or a misadjusted power steering gear may cause this condition. Tire inflation, load factors, and suspension component condition should also be inspected.

Reduced Wheel Cut: May be caused by misadjusted power steering gears, wheel stops not being adjusted properly, or bent mechanical steering components.

Wheel Kick: May be caused by air in the hydraulic system, loose steering gear mounting, worn steering linkage, front wheel bearings not adjusted or worn, improper steering gear adjustment, damaged or worn steering gear components, or worn or missing poppet valves.

Pulling and Non-recovery: May be a result of a variety of mechanical and/or hydraulic issues. In addition, front-end alignment, tire inflation, load factors, and road conditions may also have an effect on the steering components. When determining the cause of pulling and non-recovery complaints, check tire pressures, wheel bearing adjustments, suspension components, and ride height adjustments in addition to inspecting steering system operation and components. If no mechanical or hydraulic issues are found, perform a front and rear axle alignment as alignment factors may be contributing to the issue.

Fluid Leakage and Aeration: Inspect for visual signs of external and internal fluid loss. Correct as necessary. An overfilled reservoir may be the result of a restricted high-pressure line. An air leak on the low-pressure side would lead to air being ingested into the hydraulic system, which would lead to both overfill and foaming (aeration).

A loose sleeve clamp on the drag link adjuster with damaged adjusting threads will cause the steering wheel to become more off center over the course of 500 miles. A worn steering shaft U-joint or an out-of-adjustment steering gear could cause steering wheel freeplay. Defective kingpins or kingpin bearings will cause a wheel and tire not to return to the straight-ahead position during a front axle and linkage-binding test. Excessive positive caster will cause high steering effort and fast steering wheel return.

Reservoir o-rings, driveshaft seals, high-pressure fittings, and the dipstick cap are all possible leak sources. If leaks occur at any of the seal locations, replace the seal. When a leak is present at the high-pressure fitting, first tighten it to the specified torque. If this fails, replace the o-ring at this fitting.

6. Determine recommended type of power steering fluid; check level and condition; determine needed service.

Some original equipment manufacturers (OEMs) recommend the use of power steering fluid or automatic transmission fluid in power steering systems. Others recommend engine oil. Care must be taken to use the correct oil. Low fluid level will cause increased steering effort and erratic steering. It may also cause a growling or cavitation noise in the pump. Foaming in the remote reservoir may indicate air in the power steering system. Most OEM truck manufacturers recommend checking the power steering fluid level at operating or working temperature of 140–160°F.

With the engine at 1,000 rpm or less turn the steering wheel slowly and completely in each direction several times to raise the fluid temperature. Check the reservoir for foaming as a sign of aerated fluid. The fluid level in the reservoir should be at the hot-full mark on the dipstick.

7. Flush and refill power steering system; purge air from system.

Before servicing the power steering system, it is important to identify the manufacturer and to identify if the steering gear is a standard or inverted mount. Listed here are general guidelines for servicing the power steering system. Always refer to the OEM service material and follow the manufacturer's recommended procedures. When you drain and flush the power steering system, disconnect the return hose from the remote reservoir to drain the fluid. When the fluid begins to discharge from the return hose, shut off the engine. After the power steering system has been drained and refilled, the steering wheel should be turned fully in both directions with the engine running to bleed air from the system. Note: Front wheels should be raised off the ground; this will reduce heat on fluid and aid in purging air from system.

8. Perform power steering system pressure, temperature, and flow tests; determine needed repairs.

Power steering pump test procedures vary depending on the type of power steering system. Listed here is the procedure for testing power steering system pressure. Be sure to use the manufacturer's specified pressure values.

- The power steering analyzer should be connected from the power steering pump high-pressure hose to the steering gear.
- The load (gate) valve should be open at the beginning of the power steering pump pressure and flow test. Start the engine, run at idle speed, and observe the pressure and flow on the analyzer.
- If the power steering pump flow is less than the OEM specified pressure (usually 200 psi), check the high-pressure hose for restrictions. If the pressure is above the OEM specified pressure 200 psi at idle with the load valve open, check the return line for restrictions. If the flow rate is less than the OEM specification 2 gpm (gallons per minute), the pump or the flow control may be defective.
- Rotate the gate (load) valve toward a closed position until the pump pressure rises to 700 psi and increases to the test value specified in the OEM service literature. This may be between 500 and over 1,000 psi depending on the system. If the flow rate is 1 gpm less than the flow rate recorded with the gate (load) valve open, the pump or flow control valve is bad.
- Ensure that after the analyzer is removed, the engine is run, and the steering gear is turned from lock to lock and then back to center before driving the vehicle. This will ensure that no air remains in the high-pressure circuit.

9. Inspect, service, or replace power steering reservoir, including filter, seals, and gaskets.

Power steering systems may have either an integral or a remote fluid reservoir. The source of leaking power steering fluid can be the lower-sector shaft seal, submersed-style pump-to-reservoir surface, or supply-line double-flare fitting. Prying on the reservoir to adjust the pump drive belt may damage or puncture the reservoir. To remove the remote reservoir, use the following steps:

- Stop the engine, remove the cover, and, using a suction gun, remove the fluid from the reservoir.
- Use the appropriate steps to drain fluid out of the hoses and remove the reservoir.
- Remove the spring, filter cap, and filter from the bottom of the reservoir.
- Install a new filter, filter cap, and spring.
- Install the reservoir and torque to specifications.
- Fill and bleed the system.

10. Inspect and reinstall/replace power steering pump drive belts, pulleys, and tensioners; adjust drive belts and check alignment.

Loose belt tension can cause the lack of power assist in the steering system. Perform a visual inspection to determine if the belt or pulley is worn to an extent that replacement of either unit is needed. A glazed (polished, shined) belt indicates that a few items must be inspected. Belt tension checks should be performed using a belt tension gauge and checked midpoint between the pulleys at the longest belt span. Is the belt the proper length and width? If these are within specification, check the pulley for excessive wear. This can be determined by measuring the depth of a new correct belt as it sits in the pulley. Does it bottom out in the groove? If so, then the pulley is worn and must be replaced. Is the pulley damaged or loose on the shaft? Is alignment correct? This is not so critical on V-belt installations but is a constant cause of bearing failure in serpentine belt applications, as there is considerably more belt wrap and surface area involved and more side loading effect on the bearings (most of which are ball type versus tapered roller), which leads to premature failure. Manufacturers such as Caterpillar specify the use of a laser alignment tool to perform all adjustments and have "1 degree" as the ideal maximum out-of-tolerance specification between pulleys.

Note: Dual V-belt applications should use "matched set" belts to ensure consistent belt length.

11. Inspect, adjust, or replace power steering pump, drive gears/shafts, mountings, and brackets.

The power steering pump is used to produce the hydraulic circuit flow required for the power steering system. They may be belt or gear driven depending on the vehicle application. Power steering pump inspection should be performed at every PM cycle or whenever a steering complaint is issued.

When inspecting the power steering pump, consideration to be given to the pump mounting bolts to ensure that the correct torque is applied and mounting brackets checked for signs of movement, cracking, or distortion. If the pump is belt driven, the belt condition and tension should be checked. Also, the belt pulley should be inspected for wear, damage, and proper alignment.

It is necessary to test the system's pump if there is a lack of hydraulic pressure. Low pressure can be caused either by pump problems or by problems in the steering gear. Pump tests tell the technician which part is causing the problem. A complete test of the system's pump requires testing for both pressure and volume. Flow and pressure meters are installed on the complete system. After testing for both flow capacity and pressure, evaluate the situation.

When the power steering pump lacks the ability to perform its primary function properly, is low on fluid or when air is introduced into the system, a whining noise will usually be heard coming from the pump. This is accompanied by increased turning effort, binding, non-recovery, pump overheating, and hard steering.

If pressure is low and the flow is low, the pump is probably the faulty component. If pressure is low but the flow is normal, chances are that excessive internal clearance in the steering gear, due to wear, is causing the problem.

The inspection process should also include looking for external leakage. Sources for oil leakage are the drive shaft seal, reservoir o-ring seal, high-pressure outlet fitting, and the dipstick cap. Inspect the hose condition for cracking, brittleness, leakage, or chafing due to external components rubbing against the hose.

Check to see if there is an adequate amount of the correct type of fluid in the reservoir. Check the condition of the fluid, noting any discoloration. Air in the system will cause the fluid to have a foamy appearance. If the pump is overheated, the fluid will become dark. Note that if the fluid does become discolored, other problems may exist in the power steering system.

If internal damage is suspected due to low fluid levels or overheating, the power steering pump must be removed and disassembled to determine the extent of the damage. A damaged power steering pump can be identified by score marks in the pump drive gear. Elongated mounting holes in the power steering pump bracket may cause a noise while in operation. Worn holes in the power steering pump mounting bracket could cause premature belt wear. Refer to the manufacturer service literature for inspection and repair procedures.

When you replace the power steering pump drive on a truck with a diesel engine, it is not necessary to re-time the engine when this replacement process is complete. Further, you inspect the driven gear for worn or chipped teeth. You perform these tasks when removing and installing a gear-driven power steering pump: check pump mounting holes for wear, remove hoses from pump and cap the fittings, replace the o-ring, and bleed the air from the power steering system.

12. Inspect and replace power steering system cooler, lines, hoses, clamps/mountings, and fittings; check hose routing.

Power steering fluid under pressure from the pump via the high-pressure hose flows through the steering gear and continues through a low-pressure hose to the cooler and pump reservoir. You should check the power steering cooler air passages and fins for restrictions during any preventive maintenance (PM) process. You should check all power steering hoses for leaks, cracks, dents, and sharp bends. Furthermore, you should torque all pressure fittings to specifications before replacing the o-ring.

13. Inspect, adjust, or replace linkage-assist-type power steering cylinder or gear (dual system).

Linkage-assist-type power steering systems incorporate a torque valve mounted in the drag link, a power cylinder attached to the front axle assembly and tie rod cross tube, a safety valve located in the air line from the brake reservoir and the torque valve, and a manual steering gear.

Worn kingpins, improper steering-gear mesh prelude, or a bent worm gear will cause a linkage-assist-type power steering to bind when turning corners, but with short steering corrections, wheel recovery is normal. Low lubricant level in the steering gear can cause the manual steering gear assembly of a linkage-assist-type power steering to become noisy when turning the steering wheel. The mechanical components should be inspected for proper operation, adjustment, and wear whenever a steering complaint is registered.

Inspection of linkage-assist-type power steering system should start with an inspection of the air system reservoir. Drain the air tank to ensure no water or contamination is found. Inspect all air lines and fittings for chafing, proper routing, and leakage. Inspect the safety valve, the torque valve, and the power assist cylinder for proper operation, leakage, and damage. Every 25,000 miles or three months, the system should be lubricated internally. If any components are found to be in need of adjustment or are defective, refer to the manufacturer's service material for the proper procedures.

14. Inspect, adjust, repair, or replace integral-type power steering gear.

A good starting point in troubleshooting a power steering gear is checking the performance of the system in terms of both operating smoothness and steering wheel play. Begin checking the steering gear by feeling for and measuring the steering wheel freeplay with the engine running and the truck at a standstill. After determining excessive play, have an assistant turn the wheel slowly back and forth as you observe the operation of the linkages (engine on). Inspect all external steering linkage and the U-joint at the base of the steering column first for looseness.

If any part of the system is binding, the linkage may cause difficult steering and make the driver suspect power steering hydraulic problems when the trouble is a simple, mechanical part of the steering system. If the binding seems to occur in the steering gear itself, it could be due to a worn or damaged bearing. If the gear has an adjustment for the sector shaft, adjust it to specification.

Both the input and output shafts should be disconnected to measure steering gear preload torque. This process prevents binding in the steering column or linkage from affecting the worm bearing preload and sector lash adjustments. To center the steering gear, turn the worm shaft from stop to stop and count the number of turns. Starting at one end of the worm shaft travel, turn the worm shaft back half the number of turns from stop to stop. This action centers the steering gear in preparation for the sector lash adjustment.

Next, perform a steering gear internal leakage test. The first step is to prevent operation of the gear's internal unloading valves. These may be a simple pressure relief valve or a valve located at either end of the steering gear piston's travel.

Have an assistant turn the steering wheel until the steering gear contacts the axle stops. Have him or her hold it there for just a few seconds while reading pressure and flow. The pressure should be at the pump relief pressure. Flow should be compared to factory specifications for gear leakage. If steering gear leakage is above acceptable levels, the steering gear should be overhauled or replaced. Refer to OEM service literature for these procedures.

15. Adjust manual and automatic steering gear poppet/ relief valves.

When the front wheels are turned in a power steering gear, torsion bar deflection moves the spool valve inside the rotary valve. This valve movement directs power steering fluid to the appropriate side of the worm shaft piston to provide steering assist. In many steering gears, a poppet or relief valve located either in the piston or housing allows for an automatic reduction of assist pressure when the road wheels come close to, or reach, their turning limits. As piston travel approaches the steering limit, an adjustable plunger will contact and push the valve off its seat. The valve, when unseated, allows for some of the assist pressure to flow back to the fluid return side of the steering gear. This partial exhausting of pressurized fluid allows for a reduction in system temperature and pressure, relieving stress on the mechanical components of the steering system.

If the steering gear poppet/relief valves are found to be in need of adjustment, a poppet valve service kit must be installed. Consult the manufacturer's repair manual to ensure the correct procedure is being followed to install the service kit and adjust the poppet valve.

16. Inspect, align, and replace pitman arm.

The pitman arm can be directly responsible for directional stability. A damaged pitman arm can be directly responsible for the steering wheel being off center. A loose pitman arm on the steering gear output shaft could cause a steering shimmy below 30 mph. If removal or replacement of the pitman arm is required, scribe mark the pitman arm in relation to the steering gear output shaft and use a suitable puller to remove. Upon installation, align the scribe marks made upon removal. When you replace a steering or pitman arm, you should perform all of the following: road test when repairs are completed, check and correct for changes in wheel alignment, and lube the replacement part after installation.

17. Inspect, adjust, and replace drag link (relay rod) and tie rod ends (ball and adjustable socket type).

The drag link connects the pitman arm to the steering control arm. It can be either a one-piece or two-piece design. The length of the two-piece design is adjustable, which makes it easy to center the steering wheel. A loose sleeve clamp on the two-piece drag link may allow enough play in the sleeve to damage the adjusting threads, and this action may change the steering wheel position. A misadjusted drag link could cause the steering gear to operate off center and can cause a steering wheel shimmy. Before you adjust the drag link or set the toe adjustment, check ball joint motion in each end of the drag link and/or tie rod. Use hand pressure only to check ball sockets for excessive wear and movement. A worn relay rod socket or an out-of-adjustment tie rod may cause a wandering condition while driving. Also check the rubber boot for damage. Be sure to thread the tie rod end in beyond the split or damage to the relay rod may occur.

18. Inspect and replace steering arms.

A damaged or bent steering arm can cause a change in the steering wheel position and affect Ackerman geometry, turning radius, toe-out on turns, and tire wear. A damaged or bent upper steering arm in the driver's side steering knuckle can cause a change in the steering wheel position. Inspection of the steering arms should include inspection of the tie rod end mounting for looseness. If looseness is found, the tie rod end should be removed and the tapered bore checked for damage. Torque the castellated nut and install a new cotter key. To test a steering linkage joint for excessive wear you should simply push against the joint with force that you can create with your hands. This should be enough to identify excessive wear.

Depending on the type of spindle arrangement, the upper and lower steering arms may or may not be able to be replaced separate from the spindle assembly. If the upper and lower arms are replaceable, ensure that the new component matches the old component. Remove the tie rod end ball sockets using the proper tool to prevent damaging the ball socket. Once the steering arm has been replaced and the ball sockets reinstalled, torque all retaining nuts and bolts to manufacturer's specifications and be sure all cotter keys are installed and secure.

19. Inspect and replace tie rod cross tube (relay rod/centerlink), clamps, and retainers; position as needed.

The length of the tie rod determines the front wheel toe. Inspection of the cross tube should include the tie rod end ball sockets for wear, tie rod end grease boots for damage, tie rod tube for bends or damage, tie rod clamp welds, tie rod end installation, and tie rod clamp positioning. A tie rod clamp that interferes with the I-beam on a full turn will cause a noise when turning over bumps. Tie rod clamps are installed with the clamping bolt opposite the split in the tie rod cross tube. When replacing the tie rod ends, the technician must insert the threaded portion of both tie rod ends into the cross tube until they are threaded in beyond the split at each end of the cross tube.

20. Check and adjust wheel stops.

When the wheel stops are missing, damaged, or out of specified adjustment range, they can cause excessive turning radius and damage to tires, hoses, and body panels. Adjust the wheel-stop bolts to produce the correct turning radius.

B. Suspension System Diagnosis and Repair (16 Questions)

1. Inspect and replace front axle beam and mounting hardware.

Heavy-duty front axles are forged solid I-beams. The axle is mounted to the leaf springs and retained with U-bolts. When inspecting an axle the following should be considered:

- Perform a thorough visual inspection to identify looseness and shifted or broken parts.
- Inspect the U-bolts and shims for proper torque and placement.
- Measure the front wheel set back on both wheels and compare because a difference in measurement could be caused by a bent axle, or worn, misaligned components (i.e., spring eye bushings).
- Check the axle for twist using a machinist's protractor.

If the front axle must be removed, pay particular attention to spacers and castor shim placement. Be sure to mark the front of the axle to ensure proper placement when reinstalled. Also, always replace the U-bolts. U-bolt threads stretch during initial installation and scale and rust buildup may prevent improper pull-up torque and insufficient clamping force.

2. Inspect, service, adjust, and replace kingpins, bushings, locks, bearings, seals, and covers.

Some OEMs refer to the kingpin as a knuckle pin. Kingpins may be a straight or tapered design. The kingpins connect the I-beam to the steering knuckle and act as pivots to turn the front wheels. A loose kingpin or kingpin bearing could cause a shimmy with slight vibrations. A dial indicator is used to measure vertical play, upper and lower bushing freeplay, and upper and lower bushing torque deflection in the steering knuckle. Vertical end play is adjusted by adding or removing shims to obtain the manufacturer's clearance specification. If the bushing freeplay measurements are out of spec, the bushings and kingpin will need to be replaced.

When greasing a front axle kingpin, place a jack under the front axle and raise the vehicle to unload the suspension. Grease the upper bushing until grease appears between the steering knuckle and the top of the axle eye. Grease the lower bushing and continue greasing until grease begins to leak out through the pivot bearing. This shows a thorough distribution of the grease.

When replacing the kingpins, remove the upper and lower knuckle caps to access the kingpins. Remove the draw key nut and drive out the draw key. Then press out the kingpin from the knuckle. Prior to assembly, check the axle eye bore for wear.

The draw key keeps the kingpin aligned and tight in the axle bore. When reassembling the assembly, torque the draw key nut to the manufacturer's specification. The draw key torque should also be checked at every preventive maintenance (PM) inspection.

3. Inspect and replace shock absorbers, bushings, brackets, and mounts.

When a vehicle hits a bump, the wheel and suspension move upward in relation to the chassis. This causes jounce and rebound. Shock absorbers dampen or control spring action from jounce and rebound, reduce body sway, and improve directional stability and driver comfort. Worn-out shock absorbers will allow the front end to bounce, causing the steering wheel to shake for a few seconds. The piston shield often gets bent or damaged and will scrape against the shock absorber when the suspension moves up and down. Shock absorbers stop spring oscillation and are still needed on air spring suspensions.

Shock absorbers should be inspected for loose mounting bolts and worn mounting bushings. If these components are loose, rattling noise is evident, and replacement of the bushings and bolts is necessary. In some shock absorbers, the bushing is permanently mounted in the shock, and the complete unit must be replaced if the bushing is worn. When the mounting bushings are worn, the shock absorber will not provide proper spring control, and rattling will occur when driving over road irregularities.

Shock absorbers should be inspected for oil leakage. A slight oil film on the lower oil chamber is acceptable. Any indication of oil dripping is not acceptable, and unit replacement is necessary.

Shock absorbers should be inspected visually for a bent condition and severe dents or punctures. When any of these conditions are present, unit replacement is required.

A manual test may be performed on shock absorbers. When this test is performed, disconnect the lower end of the shock and move the shock up and down as rapidly as possible. A satisfactory shock absorber should offer a strong, steady resistance to movement on the entire compression and rebound strokes. The amount of resistance may be different on the compression stroke compared with the rebound stroke. If a loss of resistance is experienced during either stroke, shock replacement is essential.

Some defective shock absorbers may have internal clunking, clicking, and squawking noises, or binding conditions. When these shock absorber noises or conditions are experienced, shock absorber replacement is necessary. It is advisable to replace shock absorbers in pairs.

4. Inspect, repair, and replace (leaf and parabolic) springs, center bolts, clips, spring eye bolts and bushings, shackles, slippers, insulators, brackets, and mounts.

A leaf spring is a spring assembly where the individual leaves are the same width for the whole length of the assembly. Parabolic springs are assemblies where the leaf width is usually greater at the center of the spring, and the width decreases toward the outer end of the spring. A taperleaf spring is also classified as a parabolic spring since the leaves are thicker in the center, and the thickness decreases toward the end of the leaf. Regardless of what type of spring assembly is installed on a vehicle, the inspection process is the same.

When inspecting the leaf/parabolic spring assemblies the technician should:

- Look for broken or shifting leaves in the spring pack.

- Inspect the front and rear shackle spring pins and bushings for wear.
- Inspect the U-bolts, spring seats, and top plates for looseness or wear.

The suspension system should be inspected periodically to ensure vehicle safety. This inspection should be part of a regular maintenance schedule recommended by the truck manufacturer or truck owner. During a front or rear suspension inspection, check these components:

- Inspect the tires for excessive tread wear.
- Check the torque on the wheel nuts, and inspect the wheel rims for damage.
- On rear suspension systems, inspect all torque rods for damage or a bent condition. Inspect all torque rod bushings for looseness, wear, and deterioration.
- Inspect all spring shackles, bushings, and brackets for looseness. A pry bar may be used to pry downward on the outer end of the spring to check for looseness in the shackles and bushings.
- Inspect all spring U-bolts for damage and check the torque on the U-bolt nuts.
- Inspect the springs for broken leaves, damaged clamps, broken center bolts, and a sagged condition.
- Inspect all equalizer components for wear.

Be sure each side of the drive axle housing is positioned on the same location on each spring. A broken center bolt may allow the drive axle to slide backward on the spring.

Loose spring shackles will not break the spring center bolt. Loose front spring U-bolts can break the spring center bolt or leaves between the U-bolts. A broken leaf spring center bolt can cause axle shifting leading to premature toe-like tire wear and steering pull. Also, if the spring assembly is replaced, always replace the U-bolts. U-bolt threads stretch during initial installation and scale and rust buildup may prevent improper pull-up torque and insufficient clamping force.

If a defective spring assembly is identified, the following recommendations should be considered:

- Whenever one spring leaf is broken, all of the other leaves in the assembly have been overloaded and running under increased stress. Consideration should be made to the age and condition of the other leaves prior to repair.
- If one spring leaf is broken, and the remaining leaves are deemed serviceable, it is acceptable to replace only the broken leaf. However, it is advisable to replace the same leaf in the other spring pack on that axle to ensure both spring assemblies remain matched regarding load carrying capacity, rate of deflection, and ride height characteristics.
- If two or more leaves are broken in the spring pack, it is recommended that the entire spring assembly be replaced.
- If the spring assembly on one side of the vehicle is being replaced, it is advisable to replace both spring assemblies across that axle.

5. Inspect, adjust, and replace axle aligning devices, including radius rods/arms, torque rods, transverse torque rods/track bars, stabilizer bars, bushings, mounts, shims, and cams.

Torque rods, radius rods/arms, transverse torque rods, and torque leaves provide braking force absorption and allow for driveline angle adjustment and axle alignment.

Tandem axle suspensions often have four multi-leaf springs and four torque rods. An equalizer hanger (center hanger) is mounted to the frame between the front and rear springs on each side of the tractor. The springs ride on an equalizer pivot beam mounted on a sleeve in the equalizer bracket. Torque rods may be substituted for rear shock absorbers. Four to six torque rods may be used on spring suspensions for suspension alignment. Torque arms are either adjustable or fixed-length design. Rear axle alignment adjustment on a spring suspension with adjustable torque rods can be made by rotating one end of the torque rod in or out of the threaded sleeve. Fixed-length torque arms use shims between the front of the torque rod and spring hanger bracket to align the rear axle.

Some manufacturers utilize a multi-leaf spring pack equipped with a torque leaf. In this case, an eccentric bushing at the torque leaf is rotated to achieve axle alignment.

Torsion bars are the same length but are usually not interchangeable left to right. As long as a torsion bar is not cracked or broken, it can be adjusted at the crank assembly to level the vehicle. You never heat or bend torsion bars or torque arms.

Premature suspension bushing wear will cause noisy operation, directional instability, and excessive wear to adjacent suspension components. Inspect the torque arm/leaf bushings and pins for wear, cracking, and shifting. In addition, the technician should check the mounting bolts and shim packs on units utilizing this mounting method. When bushings are replaced, be sure the suspension is in a neutral state. A bushing not relaxed during assembly can cause a binding condition, leaving a vehicle off-level.

6. Inspect and replace walking beams, center (cross) tube, bushings, mounts, load pads, brackets, caps, and mounting hardware.

Equalizing beam suspension lowers the center of gravity of the axle load. The two types of equalizing beam suspensions are leaf spring type and rubber load cushion type. In a leaf spring-type suspension, the front of the leaf spring assembly utilizes a spring pin to mount it to the front hanger. The rear of the spring has no rigid mount allowing for forward and rearward movement to compensate for spring deflection. The spring assembly mounts to a saddle with either U-bolts or straight bolts. Tandem rear-axle suspension systems have equalizer beams on each side of the suspension. The equalizer beams incorporate bushings at each end and in the center of the beam. The equalizer beams are attached to the front- and rear-axle housings and the spring saddle. A cross tube is mounted between the two equalizer beams. You can service walking (equalizing) beam center and end bushings on the vehicle using a special OEM tool. Multi-leaf springs are mounted on saddles above the equalizer beams.

The rubber load cushion-type equalizing beam suspension fulfills the same role as the leaf spring suspension in supporting the load and absorbing road shock. The load cushions are mounted between a saddle and frame hangers. As in the leaf spring style, the axle housings and saddles attach to the equalizing beams.

When inspecting walking beam suspensions, the technician should look for beam bushing wear, spring stack shifting, loose mounting hardware, spring hanger or rubber cushion wear, and torque arms and bushings. In addition, inspect the cross tube for wear or damage. Wear in the cross tube may indicate worn equalizer beam bushings.

If a defective spring assembly is identified, the following recommendations should be considered:

- Whenever one spring leaf is broken, all of the other leaves in the assembly have been overloaded and running under increased stress. Consideration should be made to the age and condition of the other leaves prior to repair.
- If the spring assembly on one side of the vehicle is being replaced, it is advisable to replace both spring assemblies across that axle.

7. Inspect, test, and replace air suspension springs (bags), mounting plates, and main support beams/springs, pressure regulator and height control valves, linkages, lines, hoses, and fittings.

Perform a visual inspection of the pneumatic (air) system. Checks are to be performed for incomplete, binding, or improperly adjusted linkages; valves mounted in the wrong position; air line leakage, fraying, cracking, chafing, or improper positioning; or anything that could lead to future failure.

The most effective means of locating small leaks is a soap-and-water solution sprayed onto all air system components; observe bubbles as an indicator of the leak source. Test system to determine if valves are of the time delay type. Install a suitable gauge to determine the proper delivery of air pressure.

Vehicles that keep swaying back and forth or side to side can have worn-out shock absorbers, the wrong type of air valves (none or too much time delay), or insufficient air supply.

Determine proper regulated air pressure. Some systems do not operate at governed pressure.

A visual inspection will determine if the air springs and related mounting hardware are mounted in proper position so as to facilitate proper movement and rigidity. Suspension arms must be checked for bushing wear and movement. Check manufacturer's specifications for each individual system type.

Air springs (bags) should be checked for possible damage, such as cuts, exposed cords, cracking, or wear to due to misalignment. Replacement of air bags is recommended, as the cost of the new unit is economical and repair is not suggested.

The pressure protection valve should be inspected to the valve closes if the air pressure drops to less than 70 psi (482.65 kPa) to protect the air brake system from low air pressure. Some systems may incorporate a pressure regulator valve. If equipped, the pressure regulator should be checked for proper mounting, air leakage, and proper operation. Refer to manufacturer's repair manuals for proper procedures.

Leveling valves should be checked for binding, air leakage, and proper adjustment.

8. Diagnose, inspect, and replace auxiliary lift axle components and controls.

Lift axles are non-powered, and are installed ahead of or behind the driving tandem axles on a straight truck or tractor. If it is ahead of the tandem, a lift axle is called a "pusher." If it is behind, it is a "tag." Such axles can also be installed ahead of or behind a trailer's tandem. In addition, a lift axle is occasionally applied to a single-rear-axle straight truck.

Inspection of the lift axle assembly should include:

- Check all fasteners installed and bolts for proper torque specifications. Note: All fasteners torque specifications are given for dry fasteners with no additional lubrication required.
- Inspect all wheel lug nuts tightened to recommended torque specifications.
- Raise and lower the assembly and inspect the air control system for leaks and proper operation.
- As the suspension is raised and lowered, inspect for any interference between the auxiliary suspension and any truck components.

- Ensure that the brakes and slack adjusters are properly adjusted, and the wheels free to rotate.
- Check the wheel hubs for proper oil level and are filled with the manufacturer's specified lubricant.
- Check that the toe-in adjustment is set properly (1/8 +/− 1/16 measured at the tire centers).

Important: With the vehicle unloaded, the auxiliary axle's ride springs must be limited to a maximum of 20 psi to avoid improper weight distribution or component damage.

9. Measure front and rear vehicle ride heights; determine needed adjustments or repairs.

Always address the manufacturer's specifications to determine:

- The type or model of the suspension; for example, spring or air type.
- The load rating of the suspension.
- The point(s) of measurement.
- Measurement specifications.
- System air pressure and supply (if air type).
- Tools needed to perform measurements.
- How measurement will be performed; for example, flat floor, load or no load or trailer attached, or brakes applied.

Adjust as necessary to specifications.

Always perform a height setting on an air spring system with proper air pressure and supply—air dump valve closed and the system air supply valve applying air (air going into the system)—for accurate results. Perform a road test and reconfirm ride height. Improper suspension ride height will cause driveline vibrations; accelerated U-joint wear; and changes in geometry to the center of gravity (load), steering, and braking systems.

10. Verify and diagnose vehicle suspension complaints; determine needed repairs

Perform a visual inspection and road test of the vehicle to determine if the condition exists as reported. If the ride of the vehicle is unusually harsh perform the following tests:

- Test to confirm tire pressure and construction; that is, ply design radial or bias, ply count in the tread and sidewall, width of tire and rim.
- Refer to manufacturer's specifications to confirm that the recommended type of suspension for the load carried is installed on the vehicle.
- Confirm that the suspension has not been modified; that is, new springs, increase or decrease in the spring rates.
- Confirm suspension height specifications.
- Determine if alignment is within specifications (especially caster).
- Check for binding or loose steering components.
- Measure steering box and column bearing preload specifications.
- Check shock absorber function.
- Check for correct mounting of components, bushings, rods, etc.

Excessive positive caster, leaking or damaged shock absorbers, and loose or worn suspension components can cause rough ride characteristics. Improper wheel balance, worn kingpin bushings, a bent wheel mounting surface, or a shifted belt inside a tire can also cause rough ride characteristics.

11. Torque U-bolts to manufacturers' specifications.

Leaf spring suspensions require the use of U-bolts to hold the spring assembly in its proper location. Any movement between the spring and the housing can produce axle tracking problems and the possibility of shearing the spring center bolt and leaf breakage between the U-bolts. These U-bolts must be torqued to specification to produce the proper clamp load on the spring assembly to prevent any movement of the spring in relation to the axle housing. When new U-bolts are installed, they must be re-torqued after the first 1,000 miles of loaded operation and every 36,000 miles afterward. When tightening old U-bolts, ensure that no rust or thread damage is present by cleaning and inspecting the bolt and nut threads prior to performing the re-torque procedures.

12. Check axle load distribution problems on rear suspensions; check axle seat planing angles and pinion angles.

When a tandem rear axle vehicle is loaded, the load placed on each axle should be as balanced as possible. A vehicle with an axle loading imbalance will place more stress on the driveline and axle assemblies and will produce more tire wear on the axle carrying the heavier load. When the load placed on each axle varies by more than 500 lbs. (227 kg.), shims can be placed between the axle and spring of the axle with the least load. Many manufacturers have produced charts to guide the technician in selecting the appropriate shim to correct the imbalance.

When measuring the axle loading, the fifth wheel should be placed in its normal operating location and the average load that the vehicle will be carrying should be placed on the suspension. A leveled single or tandem axle scale should be used for axle measurement. If spring adjustments are required, the positioning of the axle may be altered and the drive pinion angles should be measured and adjusted if necessary to prevent the creation of any driveline vibration.

13. Inspect frame and frame members for cracks, breaks, distortion, elongated holes, looseness, and damage; determine needed repairs.

Frame sway occurs when one or both frame rails are bent inward or outward. The following are some causes: collision damage, fire damage, snowplowing abuse, and using the truck for other than original design.

Frame sag occurs when the frame rails are bent downward in relation to the rail ends. The following are some causes:

- Excessive loads
- Uneven weight distribution
- Holes drilled in frame flanges
- Too many holes drilled in the frame web
- Holes drilled too close together in the frame web
- Welding on the frame
- Cutting holes in the frame with a torch

- Cutting notches in the frame rails
- Fire
- Collision damage
- Using the truck for other than original design

Frame bow occurs when one or both frame rails are bent upward in relation to the ends of the rails. The following can cause buckle: collision, operating a dump truck with the box up and loaded, snowplow operation, unequal loading of the frame, using the truck for other than designed intent, and fire.

Diamond-shaped frame occurs when one frame rail is pushed rearward in relation to the opposite frame rail. Vehicle tracking is affected by this condition. The following causes can result in diamond-shaped frame: collision, towing, or being towed with the chain attached to only one side of the truck frame.

Frame twist occurs when the end of one frame rail is bent upward or downward in relation to the opposite frame rail. The following causes can result in a twisted frame: collision damage, rollover, rough terrain operation, and uneven loading.

Frame buckling and **mashing** is wrinkling of C-channel flanges that accompany different categories of frame damage.

Repairing frame sway, sag, bow, diamond, or twist conditions requires specialized equipment. Vehicles exhibiting any of these conditions should be taken to a frame shop having the equipment and expertise to perform the needed repairs.

After a straightening job has been completed, the technician should check the repaired vehicle carefully. This check should include an inspection of each rail to make sure that there are no buckles or wrinkles. Rear housings or trailer axles must be set at a 90-degree angle to the centerline of the chassis. The rear end should be checked to make sure that it is square with the chassis. Reinforcing sections should fit snugly in the old channels. Finally, a road test can help determine whether a frame was properly straightened.

Welding Repairs

Welding repairs on truck frames should be performed by a qualified welder. The possible exception to this is the repair of minor cracks in the frame. When making such a repair on steel frames, the shielded arc method should be used because the heat generated during welding is localized and overheating is minimized. Additional advantages are that the finished weld can be ground flush and drilled as necessary.

Welding Methods

Recommended methods for crack repair of steel frame members are shielded metal arc welding (SMAW); gas metal arc welding (GMAW), also known as metal inert gas (MIG) welding; gas tungsten arc welding (GTAW), also known as tungsten inert gas (TIG) welding; or flux cored arc welding (FCAW).

Reinforcements

Reinforcements must be made with suitable channel stock. The reinforcements should be made of a material matching the yield strength of the frame. The length of reinforcement is dependent on the type of service the truck is used in and the frequency and severity of loading. Reinforcements should be bolted rather than welded to the frame rail, keeping the bolt pattern close to the neutral fiber.

Reinforcements may be installed on the outside or the inside of the frame side rail. When installing an inner frame reinforcement at a cross member, it will be necessary to alter the length of the cross member or replace the one-piece cross member with a multi-piece (bolted) cross member. Alteration of bolt hole locations in a multi-piece cross member may be required to suit the application.

Reinforcement plates must be long enough to extend beyond the critical area so that the ends can be cut at a 45-degree angle instead of square across the frame section. If this is not possible because of obstructions, try to relieve any high-stress pressures by spreading the load over a curved section. Also avoid several holes in direct vertical alignment or holes too close together. A staggered bolt pattern with good spacing and sufficient edge distance is desirable.

Cutting the frame behind the rear axle to alter the wheelbase is usually acceptable to most manufacturers. Consult the manufacturer regarding whether reinforcement is required when extending a frame in this manner.

Bolts and Torque Specifications

Most frames today are assembled with Huck fasteners, bolts, and nuts. Others are riveted. Bolts always must be used when attaching a reinforcement. Rivets can be replaced by bolts when the frame is repaired and reinforced.

In bolted joints, the majority of the load is transferred by the clamping force between the members of the joint. Bolts must be properly torqued to develop and maintain the desired clamping force. Loose or improperly torqued bolts can lead to failure of the joint. The bolts and nuts should be inspected periodically to ensure that proper torque is maintained.

Bolts and Huck fasteners must be always replaced by fasteners that meet OEM specifications. These may be SAE Grade 5 body-bound (machined with an interference-fit shank), SAE Grade 5, or SAE Grade 8. It is poor practice to replace Grade 5 fasteners with Grade 8 because of the flex performance of each grade. Ensure that the nuts also meet OEM specification. Never consider replacing body-bound bolts with standard shank bolts because the result will be excess movement at the fastener. Body-bound bolts are machined with an interference shank designed to limit movement. Holes that are enlarged or irregularly worn may be reamed to accept the next larger bolt diameter.

If frame components are aluminum, flange-head nuts and bolts should be used, or hardened flat washers must be used next to the aluminum under both the bolt head and nut. If modification or repair requires the replacement of existing bolts with new bolts or bolts of greater length, the old nuts also should be replaced.

Drilling

Careful consideration should be given to the number, location, and sizes of frame bolt holes in the design of a vehicle. The number, location, and sizes of additional bolt holes installed to the frame subsequent to manufacture of the vehicle can adversely affect frame strength. The drilling of the frame side member presents no unusual difficulty. Standard high-speed steel drills of good quality will usually serve, provided they are sharpened properly and not overheated during sharpening or use. Because drills almost always produce a fractionally oversize (over nominal size) hole, always finish a frame hole by using a taper reamer.

Drilling Guidelines

The following are general guidelines that should be followed when drilling holes in a heavy-duty chassis frame:

- Never drill holes into the flange sections of the frame rails.
- Use existing holes whenever possible.

- Holes should be located as close to the center or neutral axis of the side member as possible.
- Maintain a minimum of 3/4 inch of material between holes.
- There should not be more than three holes located on a vertical line.
- Bolt holes should be no larger than is required for the size of the bolt being used.
- If reinforcements are used, avoid drilling holes closer than 2 inches from the ends of the reinforcement.
- Bolts must be checked periodically to ensure that the proper torque and clamping force are maintained.
- Holes are not permitted in the drop portion of the web of drop center rails.
- The center of the holes must not be closer than 1/2 inch for aluminum rails or 1 inch for steel rails to the top or bottom flange face or be less than 3 inches apart.
- Bolt holes must not be larger than those existing in the frame, such as for spring bracket bolts.
- On aluminum frame rails, chamfer both sides of all holes 0.02 inches by 45 degrees.

14. Inspect, install, or repair frame hangers, brackets, cross members, and fasteners in accordance with manufacturer's recommended procedures.

Perform a visual inspection to evaluate the condition of these components. Replace any cracked, bent, broken, or damaged components. Refer to manufacturer's service bulletins for the most up-to-date service recommendations and troubleshooting procedures.

The three common methods of attaching specialized equipment to the frame chassis are bolts, U-bolt and clamp attachments, and welding. Bear in mind that anything you attach to a frame rail is going to significantly alter the frame dynamics.

Bolt Attachments

When the holes used for fasteners are in the least critical areas possible, bolting is preferred for their versatility and strength. This type of loading stretches the lower half of the section (produces tension) and compresses the upper half of the section (produces compression). If the side rail section was without holes, the highest concentrations of stress (tension and compression) would be in the lower and upper flanges of the rail.

The uppermost hole should be located about one-third of the web depth up from the lower flange. Stress (tension) at the bottom of this hole would be approximately equal to the stress in the lower flange. Any hole located less than this distance (for example, a hole located nearer to the lower flange) will have significantly higher stress than will the flange. The reason for the difference in stress among the holes is that the upper hole is located closer to the neutral axis of the rail, or the area where the forces of tension and compression are lower. The following recommendations must be considered whenever equipment is to be bolted to the chassis frame:

- Use existing holes where possible. When holes must be drilled, they should be located no closer to the top and bottom side rail flanges than the existing holes placed by the factory.
- Avoid drilling holes in any area of the side rail web except the central one-third area. (The web is the surface area between the flanges.) Depending on loading, the top or the bottom of the side rail can be the tension side.

- Before drilling holes in the side rails, obtain approval from the OEM regional service representative.
- In all cases, avoid drilling holes in the side rail flanges.

U-Bolt and Clamp Attachment

Generally, clamping devices are the least time-consuming and least expensive methods of attachment. U-bolts avoid frame drilling, which is especially helpful when the vehicle is equipped with a heat-treated frame. But because U-bolts or clamps are not locked to the frame, there is the possibility of heavy equipment moving or being jarred out of place, especially if grease or oil is present on the frame rail. U-bolts and clamps cannot hold high-torque loads and can cause local stress concentrations.

The following guidelines must be considered when using U-bolts and clamps:

- Care should be used to block (with spacers) the side rail channel to prevent collapse of the flanges when the U-bolts are tightened. Steel blocks are preferred because wood may contract and drop out.
- The blocks used to prevent collapse of the side rail flanges should not interfere with plumbing or wiring routed along the frame rails. In addition, the blocks should not be welded to the side rail flanges.
- Because U-bolts and clamping devices depend on friction and a maintained clamping force for attachment, some bolts should directly connect the attachment to the frame side rail web to prevent the attachment from slipping.
- Do not notch frame side rail flanges in order to "force" a U-bolt fit. If the side rail flanges are too wide, obtain a larger U-bolt or use another method of attachment.

Weld Attachment

There are very few cases in which welding on a heat-treated side rail is allowable. Direct welding of the side rail flanges and web should be avoided. To avoid direct welding, equipment brackets should be welded on a separate reinforcement, and the reinforcement should be bolted to the side rail. It is recommended that before any direct welding to the side rails is performed, the manufacturer be consulted. This step could keep the warranty in effect.

15. Inspect, adjust, service, repair, or replace fifth wheel, pivot pins, bushings, locking jaw mechanisms, and mounting bolts.

The fifth wheel assembly requires periodic maintenance. The operation of the fifth wheel should be checked first while the tractor is connected to a trailer. The complete assembly should be washed and inspected for plate wear, cracks, damage, and loose or missing parts. The fifth wheel operation should be checked with a suitable lock tester for jaw or latch wear according to OEM service information. The mounting bolts and weld areas should also be checked for integrity. Any components showing wear or damage should be replaced. After all repairs are performed, fifth wheel operation should be rechecked with a suitable lock tester. Once correct operation is confirmed, the fifth wheel plate should be lubricated with water-resistant, lithium-based grease. All moving parts should also be lubricated according to manufacturer recommendations. These checks should be performed after the first 1,000 miles (1,600 km) and every 30,000 miles (50,000 km) afterward to ensure safe and secure operation.

16. Inspect, adjust, service, repair, or replace sliding fifth wheel, tracks, stops, locking systems, air cylinders, springs, lines, hoses, and controls.

A sliding fifth wheel is designed to move forward or rearward on its mounting plate. This type of fifth wheel is mounted on tracks and locked in position. The locking mechanism may be released mechanically with a lever or by air pressure supplied to an air cylinder.

Inspect the track and slider mechanisms for damaged or missing teeth and wear. The fifth wheel stop welds should be inspected to ensure weld integrity. Check the air cylinder (if equipped) for proper operation and air leakage. Inspect plunger operation and adjust if required. Lubricate all moving parts with a light, rust-resistant oil.

17. Inspect, install, repair, or replace pintle hooks and drawbars.

The kingpin stub is mounted on the trailer upper coupler assembly. Most are welded to the upper coupler assembly but both bolt-on and removable fasteners are also used. Trailer kingpins should be periodically cleaned and inspected for wear, mounting integrity to the upper coupler, and cracks.

The upper coupler assemblies are built in to the trailer assembly. The bolster plate is the lower plate in the assembly and provides the weight-bearing and pivot surface for the tractor fifth wheel plate when coupled. The upper coupler assembly should be inspected for mounting integrity, wear, cracks, plate distortion, and bowing.

The pintle hook is a device used to engage the drawbar of a towed vehicle. The locking mechanism may be mechanical, pneumatic, or hydraulic. Always make sure that the pintle hook is compatible with the drawbar eye on the tow vehicle. Most trailer pintles are made to conform to Society of Automotive Engineers (SAE) Standard J133.

Perform a visual inspection on pintle hooks and drawbars to confirm structural integrity of the components (check for cracks, welds, deformation, etc.). Install pintle hooks and drawbars of the appropriate grade of material to achieve or exceed rated loads. Repair only if manufacturer's specifications can be achieved by repair method. Replace if necessary.

C. Wheel Alignment Diagnosis, Adjustment, and Repair (13 Questions)

1. Diagnose vehicle wandering, darting, pulling, drifting, shimmy, and steering effort problems; determine needed adjustments and repairs.

Perform a vehicle road test to confirm customer complaint. Determine if the load carried is a cause of the complaint. If necessary, repeat the road test with the load in place. If load alone is not the problem, perform the following steps:

- Weigh vehicle with and without load.
- Confirm load rating for the vehicle type.
- Determine if the installed suspension is of sufficient load-carrying capacity.
- Confirm that the suspension has not been modified; that is, new springs, increase or decrease in the spring rates.

- Inspect front and rear suspension mounting, shackle pins and bushings, and torque arm pins and bushings for wear.
- Inspect vehicle for proper tire/rim type and air pressure.
- Confirm suspension height specifications.
- Check for binding or loose steering components.
- Check for frame damage; that is, twists, bends, cracks.
- Determine if alignment is within specifications (especially caster).
- Measure steering box and column bearing preload specifications.
- Check shock absorber function.

Adjust, repair, or replace all necessary components to return vehicle to the manufacturer's specifications. Repeat road test to confirm repairs and that complaint is corrected.

2. Check camber and KPI (kingpin inclination); determine needed repairs.

Camber is a setting that determines the inward or outward tilt of the top of the wheels when viewed from the front of the vehicle. It is referred to as either negative or positive camber. Moving the top of the wheels out from a neutral (vertical) position provides positive camber movement (see figure). Negative camber is the inward tilt of the wheel. A visual indication of out-of-specification camber is that the inner or outer edge of the tire shows accelerated wear in comparison to the remaining tread.

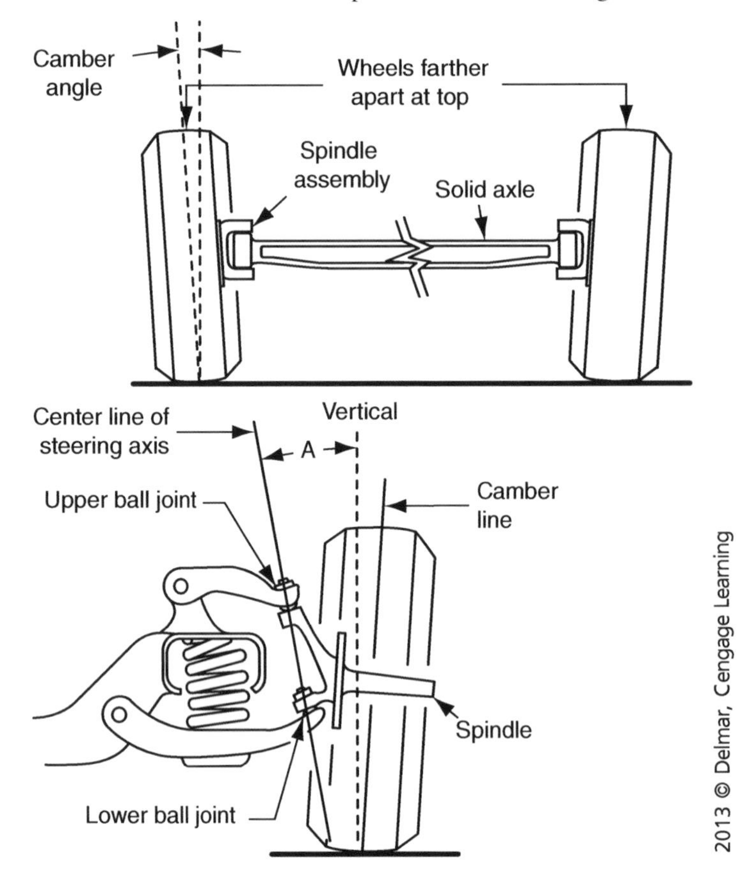

2013 © Delmar, Cengage Learning

KPI (kingpin inclination) or SAI (steering axis inclination) is the inward tilt of the kingpin at the top viewed from the front of the vehicle. KPI or SAI is measured in degrees from the center line of the ball joints or kingpin to true vertical (0). It is a directional control angle with fixed relationship to camber settings. It is also nonadjustable. One purpose of this inclination is to reduce the need for excessive camber. A road test indication of out-of-specification SAI/KPI is that the steering wheel does not return properly. Note: Check steering wheel movement, using a tension gauge first to confirm correct preload or binding.

All specifications should be confirmed using available alignment equipment after a visual/mechanical and road test inspection.

3. Check and adjust caster.

Caster is the forward or rearward tilt of the kingpin or the forward or rearward tilt of the upper ball joint in relation to the lower ball joint as viewed from the side of the vehicle. Positive caster indicates that the kingpin is tilted rearward. Determining caster settings is most accurately performed by alignment-measuring equipment and is adjusted by using angled axle shims between the front axle spring seat and front spring assembly. Caster is a necessary engineered component as it allows manufacturers to build into the vehicle directional stability. Most vehicles are designed with some degree of positive castor. Too little castor can cause wheel instability, wandering, and poor steering wheel recovery. Too much castor can result in hard steering, darting, oversteer, and low-speed shimmy (see figure).

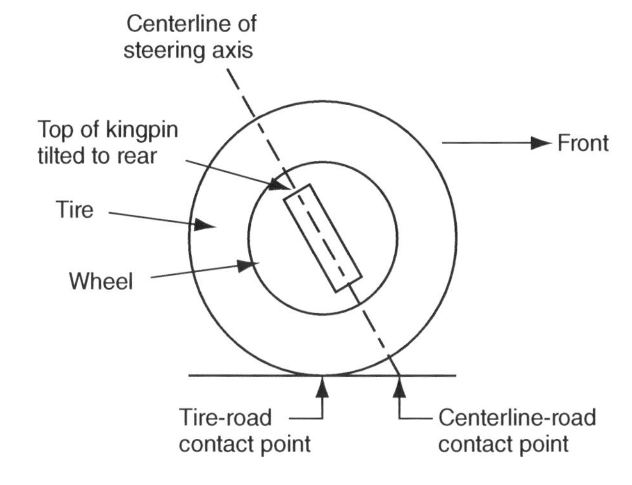

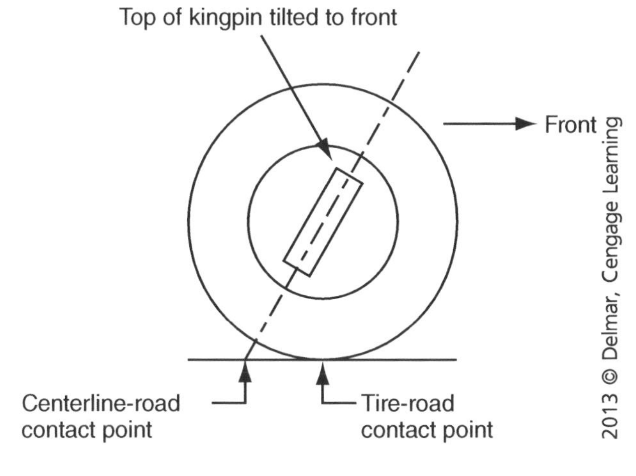

2013 © Delmar, Cengage Learning

4 Check and adjust toe.

Toe measurement can be checked by various means; usually it is performed by alignment machines, trammel bar, measuring tape, or string. A visual inspection of the tire tread area will allow the technician to determine out-of-specification settings (see figure).

Toe setting is engineered into the suspension to provide longer tire life. Vehicles having rear-wheel drive will usually have toe-in settings. As the rear drive pushes the vehicle forward the toe-in will become a zero toe measurement due to deflection in tie rods. Toe measurement is determined as follows: The front tire tread is closer together than the rear tread (toe-in) or the rear tread area is closer together (toe-out) when measured at the forward and trailing horizontal tread areas.

Incorrect toe angles not only accelerate tire wear but also can have an adverse effect on directional stability of the vehicle. Too much toe-in produces scuffing, or a featheredge, along the inner edges of the tire ribs. Excessive toe-out produces scuffing, or a featheredge, along the outer edges of the tire ribs.

Adjustment is usually achieved by rotating a threaded sleeve on the tie rod or center link or rotating the tie rod (cross tube). Although adjusting toe will not affect other alignment angles, camber settings will affect toe. If toe measurements are radically different from the OEM specifications, a measurement of the camber settings may be in order.

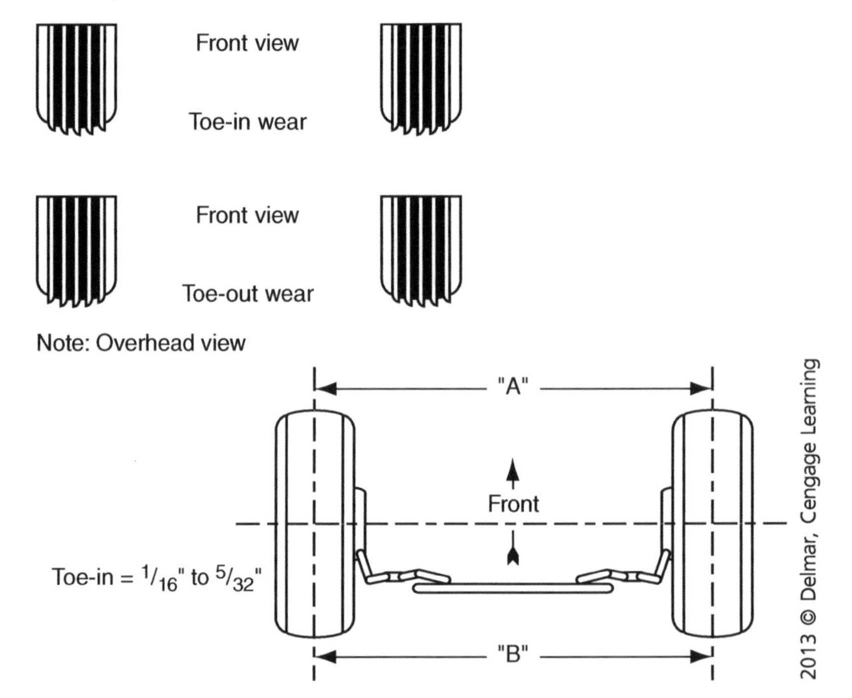

5. Check rear axle(s) alignment (thrustline/centerline) and tracking (lateral offset, parallelism); adjust or determine needed repairs.

When a vehicle's rear wheels track directly behind the front wheels, the vehicle has a proper thrustline. This prevents dog tracking or abnormal Ackerman effect. In addition, improper rear axle alignment may cause excessive scrubbing of the front steer tires.

Inspect the rear suspension and axle alignment components for wear, loose or broken parts, or distortion. Replace components as required. Adjustments to rear axle alignment may be accomplished through the use of shims or by rotating the adjustment eccentric bushings in the front spring hanger. Refer to the OEM service manuals for proper alignment procedures.

6. Check turning/Ackerman angle (toe-out-on-turns) and maximum turning radius (wheel cut); determine needed repairs.

When a vehicle turns a corner, the front and rear wheels must turn around a common center with respect to the turn radius or angle. On a single rear axle, this common center is located at the center of the rear wheels. On most front suspensions, the front wheels pivot independently at different distances from the center of the turn, and therefore the front wheels must turn at different angles. These are called the Ackerman angles, and they are controlled with the steering arms.

The inside front wheel must turn at a sharper angle compared to the outside wheel. This action is necessary because the inside wheel is actually ahead of the outside wheel (see figure).

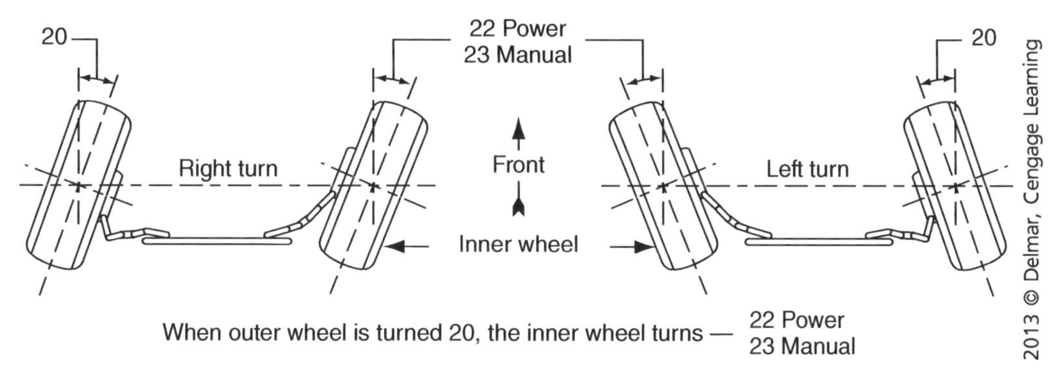

Toe-out-on-turns is checked using two turning plates equipped with angle scales and pointers. Prior to checking the Ackerman angle ensure the toe-in settings are set to OEM specifications. Mount the wheels on the turntables in a straight-ahead position. Make sure the pointers are set at 0 degrees. Turn the left-front wheel inward to a 20-degree angle and read the right wheel outside angle. Repeat the process for the right wheel and note the left wheel outside angle.

If the outer wheel angles from side to side vary more than 1 degree, further inspection is indicated. If the front-end alignment angles and toe settings are correct, and the toe-out measurement is grossly different or incorrect, one or both steering arms are bent.

Maximum turning radius is maintained by the steering stops located at each front wheel. If the steering angle is too tight, the vehicle could dangerously lock up. The technician should check the steering stop adjustment any time the front axle is replaced, replacement of the spring assemblies, or when the front-end alignment requiring a setting change. If the unit is equipped with power steering, the power steering gear pressure relief should be checked for proper setting and operation.

D. Wheels, Tires, and Hub Diagnosis and Repair (9 Questions)

1. Diagnose tire wear patterns; determine needed repairs.

To determine if repairs are needed due to premature tire wear/damage, the technician must first decide if the wear/damage was a result of external factors or vehicle component issues.

Irregular tread wear is generally a result of under/over tire inflation, component wear or alignment, and/or tire imbalance. To pinpoint the cause(s) of uneven, premature, feathering/cupping wear patterns, or internal/external damage, a visual inspection of the vehicle suspension/steering components should be performed. Next, determine if air pressure corresponds to proper tire/wheel, suspension type, and load rating. Then determine load type (weight and center of gravity) and load placement on the vehicle. This may lead to bent/ broken frame and cross members, resulting in unstable/incorrect alignment. Inside or outside edge tire wear may be contributed to improper camber angle adjustment. Featheredge wear may be contributed to improper toe settings. Cupping around the inside edge of the tire may indicate worn shocks or worn shackle bushings. Underinflated tires will wear both edges and not the center of the tire. Overinflation wears the center of the tire tread and not the edges.

To assist in the tire analysis, a technician may refer to the TMC Radial Tire Conditions Analysis Guide for a more detailed description of tire wear patterns and component diagnosis.

Note: Determine if trailer load is a factor in alignment conditions.

Return all components to manufacturer's specifications. Perform alignment measurements and set adjustments to manufacturer's specifications.

2. Diagnose wheel end vibration, shimmy, pounding, and hop (tramp) problems; determine needed repairs.

Perform a visual inspection of all vehicle tires. Check for correct air pressure and tire/wheel suspension ratings. Inspect tires/wheels for damage (i.e., bent wheels, impact bruises, out-of alignment wear patterns). Road test to confirm customer complaint. Determine if vehicle load (weight), load placement, or road condition/lane type was a factor in the complaint. Note: All of these conditions (i.e., shimmy, hop) could be caused by alignment change due to improper loading of the vehicle. Vibration is usually caused by out-of-balance or a broken tire belt. Shimmy is usually caused by a broken tire belt or worn steering component. Hop, tramp, or pounding is usually related to improper wheel balance or worn components (i.e., shock absorbers, springs/bushings, torque rods, or broken or loose frame components). Note: All of these components and conditions can combine to create interrelated symptoms.

3. Inspect and replace wheels, mounting hardware, studs, and fasteners.

Perform a visual inspection of the wheel/rim and mounting hardware including clamps, studs, and wheel nuts to determine damage or serviceability due to wear, climatic or environmental elements, or road/driver damage. Check for proper fastener grade markings. Observe for signs of overload (i.e., stretched wheel studs [perform length measurement], shifted spacers, loose nuts, or shifted rim clamps). If equipped, check if locking rings are evenly seated and secure. Locate and verify vehicle load rating and compare to actual load. Determine and follow manufacturer's installation of proper components as per manufacturer's specifications.

4. Measure wheel and tire radial and lateral runout; determine needed repairs or adjustments.

Perform runout measurements (radial and lateral) of mounted (in position) on the vehicle tire/wheel using a dial indicator with rotating wheel end. Confirm actual measurements in comparison to manufacturer's specifications. If measurements of tire runout exceed specifications, visually inspect tire/rim contact area to ascertain proper bead/rim seating. If this is correct, tire can be dismounted and the wheel can be checked for runout. Tires/wheels not conforming to specifications should be replaced.

5. Inspect tires; check and adjust air pressure to manufacturers' specifications.

Perform a visual inspection, checking for:

- Environmental damage, that is, weather checking.
- Per specifications, correct air pressure for load, suspension, tire, and rim size.
- Indicators of irregular alignment factors such as inside or outside tire tread wear or featheredging.
- Indicators of worn steering or suspension components such as tread cupping around the inner circumference of the tire or irregular wear patterns.
- Tread separation.
- Impact damage to tread, belts, sidewall, or exposed belts and cords.
- Zipper rupture, which is a fatigue failure that occurs in the upper sidewall area, beginning with the fracture of one or more ply cords, then progressing circumferentially. This is generally seen as ripples in the sidewall of the tire.
- Mismatched tire, that is, size, construction, and correct as necessary.
- Equal tire bead seating around the circumference of the rim.

6. Perform static balance of wheel and tire assembly.

Wheel balance is the equal distribution of weight in the wheel assembly with the tire mounted. This is an important factor that affects tire wear and vehicle control. With the tire/wheel unit mounted on the vehicle (brakes not applied) or balance machine, the wheel should not move from a rest position anywhere it is rotated if it is within balance statically. The static balance test indicates a heavy area in the tire or wheel. Wheel hop and a vibration felt in the vehicle are symptoms of static imbalance. Perform a visual inspection to determine that deposits of rust, rocks, ice, snow, and so on, have not become attached in any manner to the tire/wheel unit, thus causing an out-of-balance condition. Remove as necessary. Also check for proper bead seating as a mismounted tire may contribute to a static imbalance issue.

Static imbalance can be corrected by adding an offsetting amount of weight directly across (180 degrees) from the heavy area. The most common method of curing an imbalance condition is the addition of lead weights attached to the rim.

7. Perform dynamic balance of wheel and tire assembly.

Unlike static imbalance, dynamic imbalance exhibits itself as the wheel is in motion (rotating). Usually this is a result of the tire/wheel having the weight distributed unevenly from side to side. Front-end shimmy and cupping of the tire in random patterns may be caused by wheel assemblies that are either out of balance or out of round. Before performing a dynamic balance test, the following should be inspected:

- Visually inspect the tire for signs of damage (i.e., broken belt or tire not seated properly onto rim).
- Measure and correct air pressure.
- Check for proper wheel bearing endplay and correct if necessary.
- Inspect all steering and suspension components to determine if they conform to manufacturer's specifications.

Off-track wheel balancing and on-truck wheel balancing are a complementary combination for fine-tuning wheel balance. For example, a wheel vibration problem may still exist after an off-truck balance procedure. If this problem occurs, the on-truck balancer may be used to correct the problem. On-truck wheel balancers contain a drum driven by an electric motor. This drum is positioned against the tire on the vehicle, which allows the electric motor to rotate the wheel. Many on-truck balancers have a strobe light with a meter and an electronic vibration sensor. The on-truck balance procedure corrects imbalance problems in all rotating components, including brake drums or rotors.

Warning: Do not spin the wheel at excessive speeds. Spin the wheel only until the vibration occurs. Spinning the wheel at excessive speed may cause equipment or component damage or personal injury.

8. Measure tire diameter and match tires on axles.

When measuring tire diameter and matching tires, the most accurate method is to use an endless tape to measure the circumference of the tire/wheel assembly. This should be done with the tire already inflated to the proper inflation pressure and before the tire is mounted on the vehicle.

If the tires are already mounted on the vehicle, the following methods may be used:

- A tire square (similar to a carpenter's square, but larger)
- A string gauge
- A large pair of calipers
- A wooden straightedge long enough to lie across the treads of all four dual tires

Match tires on all four axle ends as closely as possible (see manufacturer's specifications). When matching tires, consideration should be given to load rating, tread design, tire construction (radial or bias), and air pressure.

When pairing tires in a dual assembly, the tire diameters should not vary more than 1/4 inch (6 mm) or the tire circumference should not vary more than 3/4 inch (19 mm). Also, in tandem drive axle applications, the total tire circumference of the first drive axle must match, as closely as possible, the total circumference of the second drive axle. If tire diameter variance in the pairs (per axle end) is not within specification, then the larger diameter tire would carry the greater load and could result in tire failure due to overheating possibilities. In addition, if the tire diameters of axle end pairs do not match, then the ABS system could log a trouble code if the variance is beyond specification. Also, the differential internal gears could become damaged. If equipped with a power divider on a tandem drive axle arrangement, the power divider differential internal gears could become damaged.

9. Remove and reinstall tire/wheel assemblies to manufacturers' specifications.

Inspect wheel/tire assembly and replace as necessary to match corresponding axle end, load rating, road/climactic conditions, and so on. Install as per recommended manufacturer's procedures, noting use of insulators, lubricant, wheel nut type, and so forth.

Be cautious when mixing tire construction or sizing. Always refer to manufacturer's specifications.

Set air pressure to recommended specifications as per load, size, and rating. Install as per recommended torque procedures. Complete road test and re-torque. Perform an additional re-torque after specified distance or time limit has occurred.

10. Clean, inspect, lubricate, and replace wheel hubs, wheel bearings, and races/cups; replace seals and wear rings; adjust wheel bearings (including one- and two-nut types) to manufacturers' specifications.

Bearings are designed to provide long life, but premature bearing failure has many causes. If a bearing fails, the technician must determine whether the failure was caused by normal wear or if the bearing failed prematurely. Premature bearing failure may be caused by:

- Lack of lubrication
- Improper type of lubrication
- Incorrect bearing adjustment
- Misalignment of related components such as shafts or housings
- Excessive bearing load
- Improper installation or service procedures
- Excessive heat
- Dirt or contamination

Clean all bearings, bearing races, and hub assemblies with clean solvent and dry with compressed air. Perform a visual inspection on bearings and race (cone) assemblies. Inspect units for wear, contamination, or overheat conditions (i.e., chrome surface worn away, blue color, pitting), bearing cage damage, or other abnormal defects.

Inspect the bearing and seal mounting surfaces on the spindle. Small metal burrs may be removed with a fine-toothed file. Seal contact areas can become pitted or worn beyond a useful condition. In most cases there are replacement repair sleeves available to compensate for this.

Install new components if original units are deemed unserviceable according to manufacturer recommendations and procedures. Note: Most bearing replacement is necessary due to contamination. This can be caused by fluid loss due to seal failure possibly caused by improper bearing preload. Perform wheel bearing adjustments as per manufacturer's recommended procedures.

11. Inspect and replace unitized hub bearing assemblies; perform initial installation and maintenance procedures to manufacturers' specifications.

The unitized hub is a permanently sealed and lubricated assembly designed to help reduce wheel-end maintenance. At regular intervals, the unitized hub should be checked for seal leaks, smooth rotation, and end-play. Rotate the hub in both directions. Check for free, smooth, and quiet rotation. If the bearings feel rough or sound noisy, the hub assembly will need to be replaced. Unitized hubs are non-serviceable in the field. Refer to the manufacturer's service manual for repair and adjustment procedures.

In addition to conventional and unitized wheel ends, PreSet® hub assemblies are also utilized by the industry. It is important to understand PreSet hub assemblies, which use pre-adjusted bearings to simplify installation, improve seal and bearing life, and reduce maintenance requirements. PreSet hub assemblies use pre-adjusted bearing technology. A more precise bearing setting is achieved by carefully controlling all critical tolerances in the hub, bearings, and the precision spacer. This level of accuracy and repeatability is nearly impossible to achieve in a shop environment using a manual bearing adjustment method.

PreSet hubs reduce the risk of either misalignment or damaged components during installation—two of the leading causes of premature seal failure.

To check the function of a PreSet hub, rotate the hub and check for free, smooth, and quiet rotation. If rotation is hampered, PreSet hubs should be serviced immediately. Unlike unitized hubs, PreSet hubs are fully serviceable. If repair is required, refer to the manufacturer's service manual.

SECTION 5

Sample Preparation Exams

INTRODUCTION

Included in this section are a series of six individual preparation exams that you can use to help determine your overall readiness to successfully pass the Suspension and Steering (T5) ASE certification exam. Located in Section 7 of this book you will find blank answer sheet forms you can use to designate your answers to each of the preparation exams. Using these blank forms will allow you to attempt each of the six individual exams multiple times without risk of viewing your prior responses.

Upon completion of each preparation exam, you can determine your exam score using the answer keys and explanations located in Section 6 of this book. Included in the explanation for each question is the specific task area being assessed by that individual question. This additional reference information may prove useful if you need to refer back to the task list located in Section 4 for additional support.

PREPARATION EXAM 1

1. A device that is used on suspensions to primarily absorb energy and dampen suspension oscillation is called:

   A. An equalizer bracket.
   B. A torque rod.
   C. A spring.
   D. A shock absorber.

2. Upon inspection, excessive play in the steering column assembly on a truck equipped with a tilt steering column is noted. Which of the following would be the LEAST LIKELY cause?

   A. Loose lock shoe in the support
   B. Faulty anti-lash spring in the centering sphere
   C. Loose tilt head pivot pins
   D. Column mounting bracket bolts loose

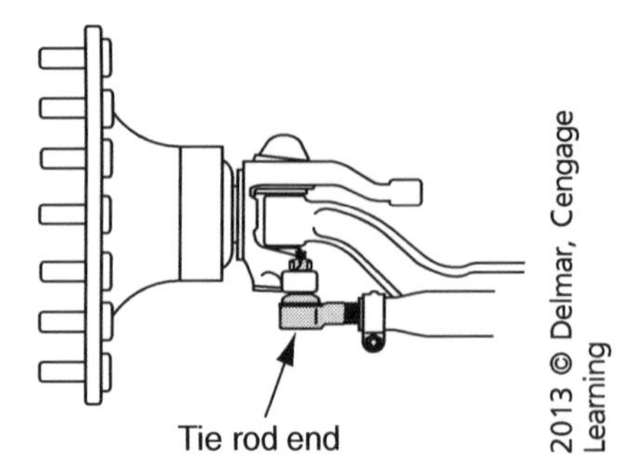

3. Referring to the figure above, when checking a tie rod end for wear which procedure is correct?

   A. Use a 2×4 to pry on the cross tube.
   B. Use a dial indicator and hydraulic jack.
   C. Use up and down hand force to check for wear.
   D. Use a feeler gauge to check for wear.

4. While discussing steering system components, Technician A states that a typical single-axle truck system has one Ackerman arm. Technician B says that the Ackerman arm controls turning radius. Who is correct?

   A. A only
   B. B only
   C. Both A and B
   D. Neither A nor B

5. A driver complains that after hitting a bump his vehicle suddenly veers to the right or left. Which of these is the LEAST LIKELY cause?

   A. A loose idler arm
   B. A damaged relay rod
   C. A worn tie rod end
   D. A wheel out of balance

6. The alignment angle that is LEAST LIKELY to cause the greatest tire wear is:

   A. Castor.
   B. Turning radius.
   C. Toe-in.
   D. Camber.

7. Technician A states that some steering shaft universal joints have different splines on the upper and lower yokes. Technician B says that when installing the universal joint, install it on the lower end of the intermediate shaft first. Who is correct?

   A. A only
   B. B only
   C. Both A and B
   D. Neither A nor B

8. Technician A states that in order for a steerable suspension to steer or track correctly, it is necessary for the front wheels to be in a toe-out condition. Technician B says that the toe setting should be between 1/16 inch and 1/8 inch. Who is correct?

   A. A only
   B. B only
   C. Both A and B
   D. Neither A nor B

9. Alignment procedures are being discussed. Technician A says that installing the caster shim between the axle and spring with the thick part to the rear will make the caster angle positive. Technician B says installing the caster shim between the axle and spring with the low side at the front will make the caster angle more negative. Who is correct?

   A. A only
   B. B only
   C. Both A and B
   D. Neither A nor B

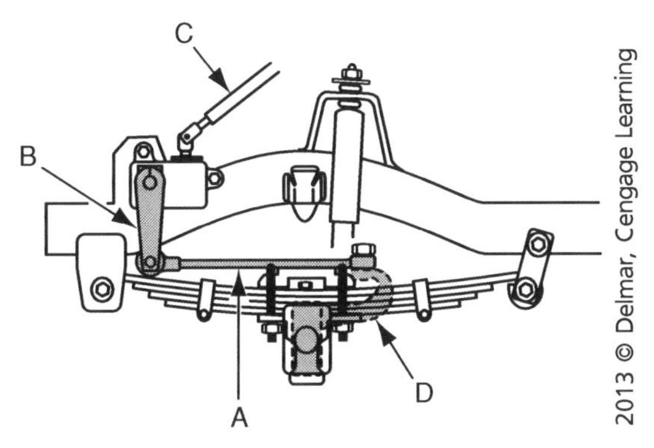

10. Referring to the figure above, the drag link is:

    A. Letter A.
    B. Letter B.
    C. Letter C.
    D. Letter D.

11. A driver complains about rough ride. Which of these would be the LEAST LIKELY cause?

    A. Excessive positive caster
    B. Leaking or damaged shock absorber
    C. Loose/worn suspension component
    D. Defective rear wheel bearings

12. When a technician is going to check out the front-end geometry of a heavy-duty truck, which of the following steps should be performed first?

    A. Neutralize the suspension by driving the unit back and forth in a straight line.
    B. Jack the front end up and scribe the front tires.
    C. Disconnect the drag link.
    D. Check the over-center adjustment of the steering gear.

13. A driver complains of excessive play in the steering column assembly on a truck equipped with a tilt steering column. All of the following might be the cause EXCEPT:

    A. Loose column mounting bracket bolts.
    B. Loose tilt head pivot pins.
    C. A loose lock shoe pin in the support.
    D. A faulty anti-lash spring in the centering sphere.

14. A truck tire is found to have been run in an underinflated condition. Technician A says the tire may be damaged internally and should be removed from the truck and inspected. Technician B says that a tire should be balanced prior to checking wheel runout. Who is correct?

    A. A only
    B. B only
    C. Both A and B
    D. Neither A nor B

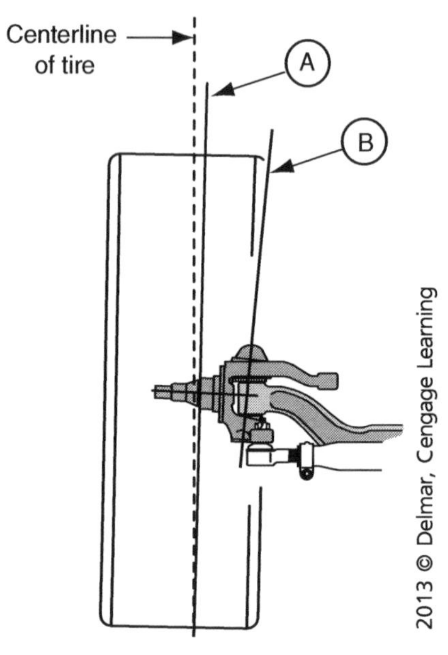

15. Referring to the figure above, what angle is being represented at Letter B?

    A. Vertical angle
    B. Kingpin inclination
    C. Camber
    D. Caster

16. If the power steering gear poppet valves are misadjusted, which of the following conditions may be exhibited?

    A. Steering wheel kick
    B. Reduced wheelcut
    C. Directional pull
    D. Non-recovery

17. A vehicle is found to be "dog tracking." Upon inspection, the technician finds no issues with worn components and says the axle alignment should be checked. Which is the LEAST LIKELY method to use for checking axle alignment?

    A. Light and laser beam systems
    B. Computer-controlled sensor systems
    C. Straightedge and tram method
    D. String and stick method

18. Technician A says excessive load/weight on a steer axle could not cause abnormal tire wear or steering issues. Technician B says excessive load/weight on one wheel end of the steer axis could cause abnormal tire wear or steering issues. Who is correct?

    A. A only
    B. B only
    C. Both A and B
    D. Neither A nor B

19. Which of the following is the acceptable method for removing a steering wheel from the steering shaft?

    A. Use a steering wheel puller.
    B. Firmly grasp with both hands and pull.
    C. Use two pry bars and remove.
    D. Strike the center shaft with a hammer to drive it out of the steering wheel.

20. When performing kingpin bushing wear dial indicator checks, what is the wear limit?

    A. 0.010 inches
    B. 0.100 inches
    C. 0.030 inches
    D. 0.025 inches

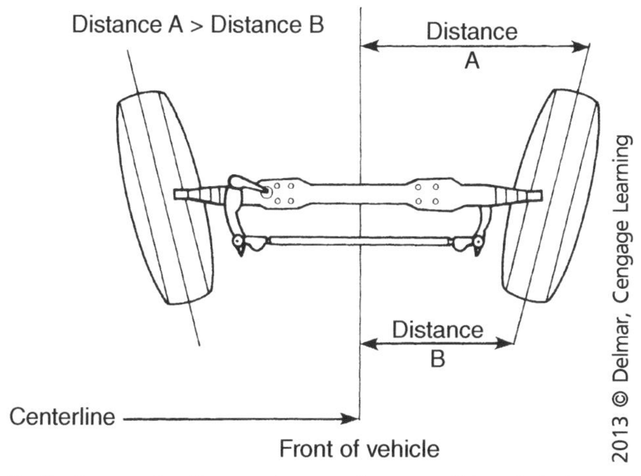

21. Referring to the figure above, what condition is being represented?

    A. Toe-in condition
    B. Camber condition
    C. Toe-out condition
    D. Steering axis inclination

22. Technician A says when aligning the rear axle position on some spring designs the upper and lower torque rod collars are rotated. Technician B says you align some rear axles by rotating eccentric bushings in the spring hangers. Who is correct?

    A. A only
    B. B only
    C. Both A and B
    D. Neither A nor B

23. A Technician has found a tire that is under 80 percent of the recommended inflation pressure. Technician A says that before removing the tire and wheel assembly from the vehicle, the technician should deflate the tire by removing the valve core completely. Technician B says that once the tire is deflated and removed it should be placed in a restraining device to check for leaks. Who is correct?

    A. A only
    B. B only
    C. Both A and B
    D. Neither A nor B

24. A vehicle leans to one side. Which of these is the LEAST LIKELY cause?

    A. One or more broken leaf springs
    B. A weak or fatigued spring assembly
    C. Unmatched spring design/load capacity spring assemblies
    D. A bent or twisted frame rail

25. Two technicians are discussing the procedure for diagnosing a steering system using a steering system analyzer. Which of the following should be true to obtain accurate results?

    A. The engine oil should be at operating temperature.
    B. The steering system hydraulic oil should be cold.
    C. New steering system hydraulic oil should be used.
    D. The steering system hydraulic oil should be above 140°F (60°C).

26. While discussing air suspension height control valves, Technician A says that an improperly adjusted air suspension height control valve can cause an offset in vehicle attitude. Technician B says that before adjusting the height control valve, the air system pressure needs to be above 80 psi. Who is correct?

    A. A only
    B. B only
    C. Both A and B
    D. Neither A nor B

27. Technician A says packing a wheel bearing with grease before filling the hub assembly with oil is acceptable because it provides the wheel bearing with lubricant while the oil settles and has a chance to run into the bearing. Technician B says this is acceptable because the oil and grease will mix once the hub reaches operating temperature. Who is correct?

    A. A only
    B. B only
    C. Both A and B
    D. Neither A nor B

28. A bent steering arm was replaced and now the vehicle produces excessive tire side scrub during sharp turns. What could cause this condition?

    A. An incorrectly adjusted camber setting
    B. A lack of kingpin lubrication
    C. An incorrect steering arm
    D. An incorrect toe setting

29. A driver complains of a front end shimmy with slight vibrations. Technician A says loose or worn kingpins or kingpin bushings might be the cause. Technician B says that an overloaded trailer might be the cause. Who is correct?

    A. A only
    B. B only
    C. Both A and B
    D. Neither A nor B

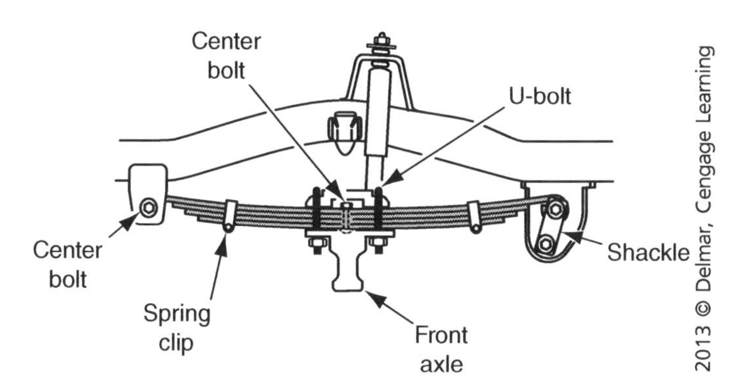

30. Referring to the figure above, what component provides self-dampening properties to the leaves in a spring pack?

    A. Spring clips
    B. U-bolts
    C. Center bolt
    D. Spring pins

31. A technician is inspecting a hub assembly and finds two broken studs. The technician should:

    A. Remove and replace the broken studs plus the adjacent studs.
    B. Remove the broken studs plus half of the remaining studs.
    C. Remove and replace all of the wheel studs.
    D. Only replace the two broken studs.

32. All of the following can cause abnormal tire wear, shimmy or vibration EXCEPT:

    A. Tire/wheel imbalance.
    B. Excessive wheel or hub runout.
    C. Excessive cam brake stroke.
    D. Improper tire mounting.

33. Upon inspection, a technician finds a single leaf broken within the spring pack. Which of the following is the LEAST LIKELY cause?

    A. The technician would replace only the broken leaf.
    B. The technician would make a careful inspection of the other leaves in that spring pack.
    C. The technician would replace the broken leaf and the same leaf in the other spring assembly on that axle.
    D. The technician would replace both the spring pack with the broken leaf and the other spring pack on that axle.

34. A driver complains that he notices a vibration in the steering wheel while traveling at 28 mph. Of the conditions listed below, which is the LEAST LIKELY cause?

    A. Improper wheel balance
    B. Worn kingpin bushings
    C. Bent wheel mounting surface
    D. Shifted belt inside tire

35. All of the following are a function of the torque arm EXCEPT:

    A. To retain axle alignment.
    B. To control axle torque.
    C. To control ride height.
    D. To adjust axle alignment.

36. While discussing wheel balance, Technician A says that dynamic wheel balance is better than static balance because it is performed with the wheel stationary. Technician B says that dynamic wheel balance is better than static balance because you can tell if the wheel is bent. Who is correct?

    A. A only
    B. B only
    C. Both A and B
    D. Neither A nor B

37. Technician A says that when reinstalling rubber-bushed equalizing beams, you should install the torque rods before torquing the beam end bolts to specification. Technician B says that you should ensure the beams are parallel to the frame rails. Who is correct?

    A. A only
    B. B only
    C. Both A and B
    D. Neither A nor B

38. Which of the following is the most likely cause of steering wheel shimmy?

    A. A toe-in setting that is too high
    B. Improper dynamic wheel balance
    C. An excessive load in the vehicle
    D. A toe-out setting that is too high

39. A technician has checked the pulley alignment on a power steering system and found them to be misaligned. All of the following could cause the power steering pump pulley to become misaligned EXCEPT:

    A. An overpressed pulley.
    B. Loose fit from pulley hub to pump shaft.
    C. A worn or loose pump mounting bracket.
    D. A broken engine mount.

40. The unitized wheel end may be described as:

    A. Easily identifiable by the two distinct star shapes embossed on the center of its hubcaps.
    B. An enclosed unit with bearings that will need to be serviced every 100,000 miles.
    C. An enclosed unit with bearings that will need to be serviced every 500,000 miles.
    D. An enclosed unit with bearings lubricated for the life of the hub, bearing, and seal assembly.

41. A truck that is equipped with an air suspension system has one side rising after unloading. Technician A says that this could be caused by a malfunctioning leveling valve. Technician B says that this could be caused by a plugged exhaust port. Who is correct?

    A. A only
    B. B only
    C. Both A and B
    D. Neither A nor B

42. All of the following can cause hard steering EXCEPT:

    A. A dry fifth wheel.
    B. An overloaded steer axle.
    C. A contaminated power steering system.
    D. Leaking shock absorbers.

43. Two technicians are discussing air cab suspensions. Which of the following statements is correct?

    A. It has three springs.
    B. It has a leveling valve.
    C. It is used with rubber cab mounts.
    D. It does not have shock absorbers.

44. Technician A says a steer axle vibration or shimmy could be caused by steer tire or wheel issues. Technician B says replacing the front tires/wheels from another vehicle with no vibration or shimmy is a viable diagnostic procedure. Who is correct?

    A. A only
    B. B only
    C. Both A and B
    D. Neither A nor B

45. During a front axle assembly inspection, a technician should check for a twisted axle beam if any of the following conditions exist EXCEPT:

    A. The difference in caster angle exceeds 0.5 degrees from side to side.

    B. The caster shims in place differ by 1 degree or more.

    C. A low-speed shimmy exists, and there is no evidence of looseness elsewhere in the steering system.

    D. Excessive tire wear exists.

46. During a preventive maintenance inspection, the technician finds that one spring pin will not take grease. The first thing the technician should do is:

    A. Change the zerk fitting.

    B. Remove the weight from the spring assembly.

    C. Replace the spring pin and bushing.

    D. Heat the spring eye.

47. Technician A says that before checking the fluid level in the power steering reservoir, you should turn the steering wheel slowly and completely from side to side several times to boost the fluid temperature. Technician B says that foaming in the reservoir indicates low fluid level or air in the system. Who is correct?

    A. A only

    B. B only

    C. Both A and B

    D. Neither A nor B

48. Kingpin inclination (KPI) may be defined as:

    A. The forward or rearward tilt of the kingpin at the top.

    B. The tracking angle of the tires from a true straight-ahead track.

    C. The inward or outward tilt of the top of the wheel when viewed from the front of the vehicle.

    D. The amount in degrees that the top of the kingpin inclines away from vertical as viewed from the front of the vehicle.

49. Upon inspection, a technician finds one air bag blown out on a tractor with an air ride suspension. The technician inspects the rest of the suspension system and finds no mechanical or fatigue issues. Technician A says that it is acceptable to replace just that air bag. Technician B says that the axle alignment should be checked after replacement. Who is correct?

    A. A only

    B. B only

    C. Both A and B

    D. Neither A nor B

50. A technician found the splines on the pitman arm worn. After replacement, the technician should do all of the following EXCEPT:

    A. Perform a road test when repairs are completed.

    B. Replace both outer tie rod ends.

    C. Check and correct changes in wheel alignment.

    D. Lube the replacement part after installation.

PREPARATION EXAM 2

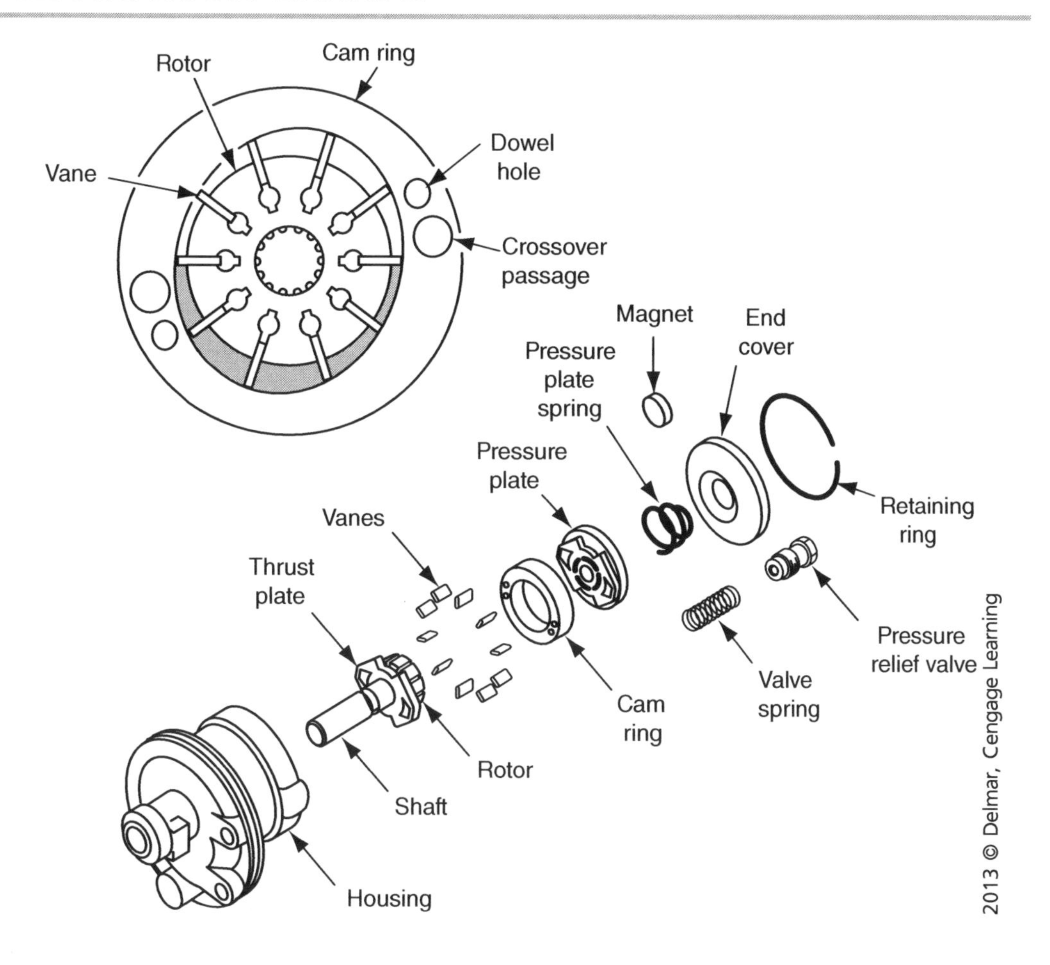

1. Referring to the figure above, a vane-type power steering pump is being tested. When using a power steering system analyzer with the engine running, the pressure gauge reads low system pressure with the shutoff (load) valve closed. Technician A says the pressure relief valve may be frozen open. Technician B says the pump vanes may be sticking in their slots. Who is correct?

   A. A only
   B. B only
   C. Both A and B
   D. Neither A nor B

2. The most common type of power steering hydraulic pump used on most Class 8 trucks/ tractors equipped with power steering systems is:

   A. Gear.
   B. Vane.
   C. Plunger.
   D. Centrifugal.

3. Technician A says to inflate mounted tires in a safety cage or using a portable lock ring guard. Technician B says to first mount the tire on the truck, then inflate to the proper tire pressure. Who is right?

   A. A only
   B. B only
   C. Both A and B
   D. Neither A nor B

4. While performing an inspection on a heavy-duty tractor, the technician finds the frame rail buckled. Which of the following might be the cause for this condition?

   A. Too many holes drilled in the frame
   B. Extreme operating conditions
   C. The wrong bolts used in a repair
   D. A vehicle collision

5. Camber may be defined as:

   A. The forward or rearward tilt of the kingpin at the top.
   B. The tracking angle of the tires from a true straight-ahead track.
   C. The inward or outward tilt of the top of the wheel when viewed from the front of the vehicle.
   D. The amount in degrees that the top of the kingpin inclines away from vertical as viewed from the front of the vehicle.

6. A technician finds that the power steering system is overheating. This condition may be caused by all of the following EXCEPT:

   A. A loose steering shaft flex joint.
   B. Underlubricated ball joints.
   C. A kink or pinch in the fluid return line.
   D. Blocked airflow across the heat exchanger.

7. Suspension components are replaced on a medium-duty truck with single-air-valve rear air suspension. Technician A says a tape measure can be used to measure from a frame straightedge to the drive axle. Technician B says if the distance from the axle to a fixed straightedge is different on each side of the suspension, one of the front equalizing beam bushings may be worn. Who is correct?

   A. A only
   B. B only
   C. Both A and B
   D. Neither A nor B

8. Technician A says that the minimum and maximum air pressure going to the air bags is controlled by the pressure protection valve. Technician B says that the minimum air pressure in the bags should not to be less than 3 psi. Who is correct?

   A. A only
   B. B only
   C. Both A and B
   D. Neither A nor B

9. Excessive tire wear is noted on a heavy-duty tractor. Which of the following suspension components is the LEAST LIKELY cause?

   A. Bent spindle

   B. Improper tie rod setting

   C. Center cross tube bushing

   D. A preloaded axle wheel bearing

10. When discussing poppet relief valve adjustment, a technician turns the steering wheel to the left until the steering effort increases. He then measures the distance between the left stop screw and the axle stop. The clearance is found to be 1/4 inch. After checking the stop screw adjustment, the technician finds that the stop screw is within adjustment specifications. The technician should:

    A. Adjust the stop screw to maintain a 1/8-inch clearance.

    B. Adjust the stop screw to a clearance of 3/8 inch.

    C. Adjust the poppet valves until the clearance is 1/8 inch.

    D. Do nothing. The clearance is okay.

11. When performing an axle alignment on a spring suspension with torque rods, all of the following apply EXCEPT:

    A. The adjustment is made through a lower adjustable torque rod.

    B. The shims are used between the torque rod front and spring hanger bracket.

    C. The adjustment is made with an eccentric bushing at the torque leaf.

    D. The adjustment is made through the upper adjustable torque rods.

12. Two Technicians are talking about inspection of collapsible steering columns. Technician A says that a dial indicator stem should be placed against the lower end of the steering shaft and then the steering wheel should be rotated. If the runout on the dial indicator exceeds the OEM specification, the steering shaft is bent and must be replaced. Technician B says that if the steering shaft is not bent but shows sheared injection plastic, the shaft must be replaced. Who is correct?

    A. A only

    B. B only

    C. Both A and B

    D. Neither A nor B

13. A driver complains that the front end hops, and upon inspection a technician notes that there are cupping marks around the inner circumference of the tire. Which of these is the LEAST LIKELY cause?

    A. Shock absorbers

    B. Worn spring shackle pins and bushings

    C. Static wheel imbalance

    D. Camber angle out of spec

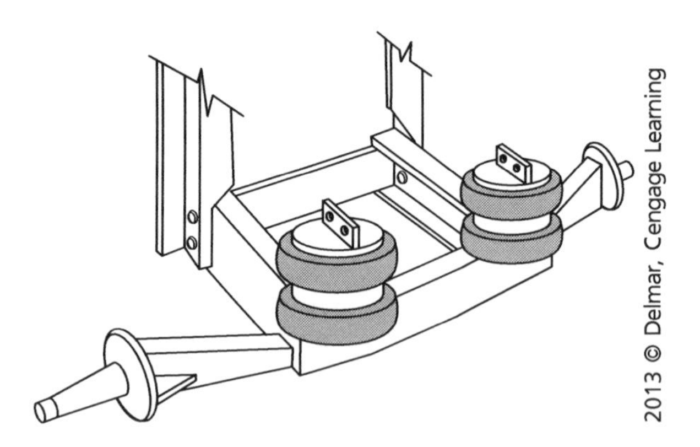

14. Referring to the figure above, while discussing the lift axle, Technician A states that the lift controls may be mounted in the cab or mounted externally. Technician B says that when raising or lowering the lift, system air pressure may drop. Who is correct?

    A. A only
    B. B only
    C. Both A and B
    D. Neither A nor B

15. During an inspection, you see wetness on the shock body. What is your next step?

    A. Perform a shock leak test.
    B. Replace the wet shock.
    C. Replace all shocks.
    D. Determine if the shock is actually leaking.

16. Toe-out on turns is defined by what component?

    A. Pitman arm
    B. Ackerman arm
    C. Drag link
    D. Steering control arm

17. Of the following, which is the LEAST LIKELY method used to adjust the axle alignment on a spring suspension with torque rods?

    A. Adjustment is made through the lower adjustable torque rod.
    B. Shims are used between the torque rod front and spring hanger bracket.
    C. Adjustment is made with an eccentric bushing at the torque leaf.
    D. Adjustment is made through the upper adjustable torque rod.

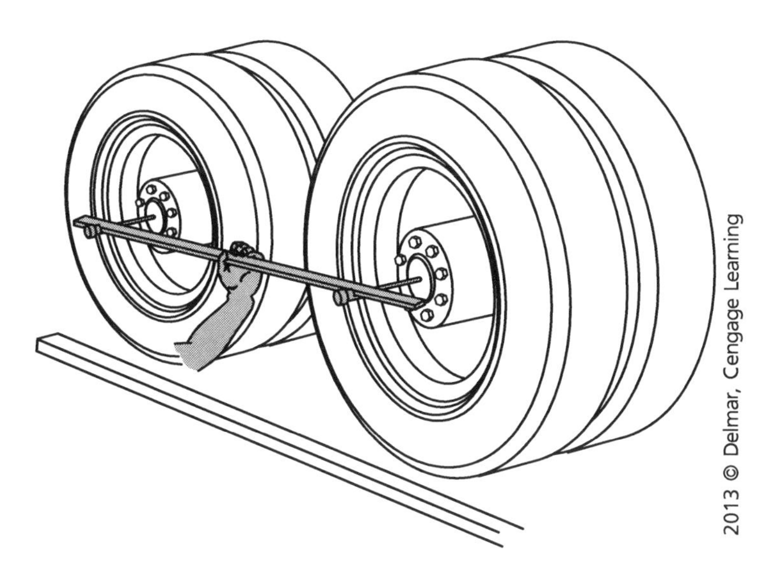
2013 © Delmar, Cengage Learning

18. Referring to the figure above, what is being checked by the technician in the figure?
    A. Wheel-bearing end-play
    B. Tandem axle spread
    C. Rear toe-in
    D. Rear wheel runout

19. Which of the following transfers the trailer weight to the fifth wheel?
    A. The bolster plate
    B. The kingpin
    C. The pintle hook
    D. The drawbar

20. Missing, damaged, or out-of-specified adjustment wheel stops may cause:
    A. Wheel imbalance.
    B. Excessive turning radius.
    C. Steering wheel to be off center.
    D. Steering wheel nibble on turns.

21. Technician A says a badly worn tie rod end can cause steering wander. Technician B says you check tie rod end wear by applying hand pressure as close to the socket as possible. Who is correct?
    A. A only
    B. B only
    C. Both A and B
    D. Neither A nor B

22. While inspecting the power steering system on a tractor, the fluid is found to be discolored. All of the following could cause this EXCEPT:
    A. The wrong type of fluid.
    B. Mixed brands of the recommended fluid.
    C. Water mixed with fluid.
    D. An overheated condition.

23. When using a trammel gauge to perform an axle alignment on a vehicle equipped with a tandem axle, Technician A states that the front drive axle should be aligned to the frame. Technician B says that the rear drive axle should be aligned to the frame. Who is correct?

    A. A only
    B. B only
    C. Both A and B
    D. Neither A nor B

24. During a preventive maintenance inspection, the power steering pump mounting bracket bolts are found to be loose. Technician A says elongated mounting holes in the power steering pump bracket may cause a noise while in operation. Technician B says worn holes in the power steering pump mounting bracket could cause premature belt wear. Who is correct?

    A. A only
    B. B only
    C. Both A and B
    D. Neither A nor B

25. What is the LEAST LIKELY purpose for using a tire restraining device?

    A. To protect the technician during tire inflation
    B. To inflate a tire that has been damaged to check for leaks
    C. To contain the tire and rim parts in the event of an explosion
    D. To hold the tire upright in order to measure its circumference

26. All of the following can cause a front steer axle pulling condition EXCEPT:

    A. A dragging brake.
    B. An out-of-adjustment brake.
    C. Incorrect brake timing.
    D. An incorrect crack pressure relay valve.

27. All of the following are functions of the cross-members of a chassis frame EXCEPT:

    A. To control axial rotation and longitudinal motion of the rails.
    B. To protect wires and tubing that are routed from one side of the vehicle to the other.
    C. To reduce torsional stress transmitted from one rail to the other.
    D. To provide a mounting surface for the fifth wheel assembly.

28. When adjusting the toe setting on a truck equipped with radial steering tires, which of the following specifications would apply?

    A. 1/16 inch
    B. 1/8 inch
    C. 3/16 inch
    D. 1/4 inch

29. When adjusting vehicle ride height, which of the following applies?
    A. The tractor suspension must be in a laden position with a full load applied.
    B. The air system pressure must be at 100 psi.
    C. Use safety stands with a sufficient load rating to support the vehicle.
    D. Move the height control valve in an upward motion to deflate the air bags.

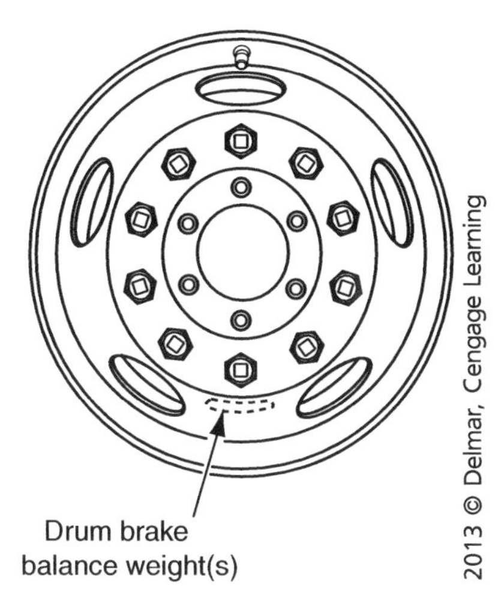

30. Referring to the figure above, a technician is performing a static balance procedure on a truck wheel. Technician A says the maximum wheel weight per tire should not exceed 18 ounces. Technician B says if 16 ounces of weight is required at one spot, use an 8-ounce weight on each side of the rim directly across from each other. Who is correct?
    A. A only
    B. B only
    C. Both A and B
    D. Neither A nor B

31. A vehicle suddenly veers to the right or left after striking a bump with the front wheels. Which of these is the LEAST LIKELY cause?
    A. A loose idler arm
    B. A damaged relay rod
    C. A worn tie rod end
    D. A wheel that is out of balance

32. After the front springs were replaced on his tractor, a driver complains that it requires excessive steering effort while turning and that the steering wheel return is too fast. Which of these is the most likely cause?
    A. Incorrect turning angle
    B. Too much negative camber
    C. Too much positive caster
    D. Underinflated tires

33. Technician A says that the pitman arm converts output torque from the steering gear into the control force applied to the drag link. Technician B says that when removing the pitman arm, scribe both the pitman arm and the steering gear shaft prior to removal. Who is correct?

    A. A only
    B. B only
    C. Both A and B
    D. Neither A nor B

34. All of the following are a result of loose U-bolts EXCEPT:

    A. Axle seat damage.
    B. A broken center bolt.
    C. Leaf spring breakage between the U-bolts.
    D. Leaf spring breakage outside of the U-bolts.

35. The two main designs of truck wheels are:

    A. Single piece and split side rims.
    B. Aluminum and hub-piloted.
    C. Disc and drum.
    D. Hub and spoke.

36. While discussing leaf spring packs, Technician A states that because leaf springs are clamped together with some force, any movement of the assembly must first overcome friction between the leaves. Technician B states that this condition is called interleaf friction. Who is correct?

    A. A only
    B. B only
    C. Both A and B
    D. Neither A nor B

37. All of the following are true statements regarding walking beams EXCEPT:

    A. Beams can be constructed of aluminum.
    B. Beams can be constructed of cast steel.
    C. Beams can be constructed of nodular iron.
    D. All manufactured beams can be stressed in either direction.

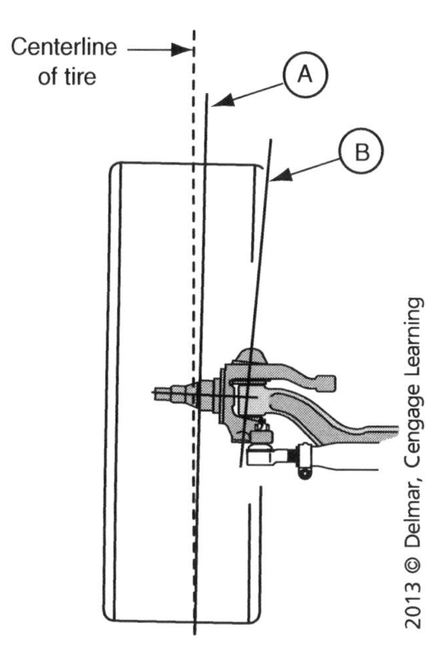

38. Referring to the figure above, what angle is being represented at Letter A?

    A. Vertical angle
    B. Kingpin inclination
    C. Camber
    D. Caster

39. After removing the tire from the rim, the technician notices that the rim is cracked. The technician should:

    A. Carefully weld the cracked wheel to repair it.
    B. Use a brazing rod to rework the wheel and return it for use.
    C. Discard it in a pile of other rims.
    D. Mark the wheel as unserviceable and remove it from the area.

40. When checking the power steering fluid level, most original equipment manufacturers (OEMs) recommend not checking the fluid level until the system reaches an operating temperature of:

    A. 100°F.
    B. 240°F.
    C. 90°F.
    D. 175°F.

41. Technician A says that a machinist protractor may be used to check the caster angle. Technician B says that sometimes, when checking for a twist in the axle, the U-bolts will need to be loosened to relieve axle tension. Who is correct?

    A. A only
    B. B only
    C. Both A and B
    D. Neither A nor B

42. You are working with a four-piece nut system with PreSet hub assemblies and the inner nut is torqued to 300 ft-lbs. If the lock ring or spindle washer does not line up with the dowel on the inner nut you should:

    A. Back the inner nut off until the lock ring or spindle washer hole lines up with the dowel.
    B. Back the inner nut off to 250 ft-lbs and install the lock ring or spindle washer.
    C. Advance the inner nut until the lock ring or spindle nut hole lines up with the dowel.
    D. Back the inner nut off to 50 ft-lbs and install the lock ring or spindle washer.

43. Of the following, which would be the LEAST LIKELY source of leaking power steering fluid?

    A. Lower sector shaft seal
    B. Submersed style pump-to-reservoir surface
    C. Supply line double-flare fitting
    D. Steering gear input shaft

44. After a front alignment has been performed, a driver complains of a pull. Technician A says that the tractor will pull to the side with the most positive caster angle setting. Technician B says that the tractor will pull to the side with the most positive camber setting. Who is correct?

    A. A only
    B. B only
    C. Both A and B
    D. Neither A nor B

45. During an inspection, a Technician finds a crack in the cross-member web that is extending into the flange. While preparing to repair a frame cross-member, Technician A says to disconnect the truck batteries before welding a frame. Technician B says it is permissible to weld across frame flanges. Who is correct?

    A. A only
    B. B only
    C. Both A and B
    D. Neither A nor B

46. Technician A says noise from the steering gear might be caused by misalignment of the steering column input shaft. Technician B says that noise coming from the manual steering gear assembly of a linkage-assist-type power steering when turning the steering may be caused by low lubricant level. Who is correct?

    A. A only
    B. B only
    C. Both A and B
    D. Neither A nor B

47. A driver complains that while going down the road the vehicle bounces excessively and pulls to the right. Technician A says the first thing that should be done is to replace the front shock absorbers and align the front end. Technician B says the first thing that should be done is to balance the front wheels. Who is correct?

    A. A only
    B. B only
    C. Both A and B
    D. Neither A nor B

48. On a typical single-axle truck system, the Ackerman arms are linked by what component?

    A. A drag link
    B. A pitman arm
    C. A steering column
    D. A tie rod assembly

49. A truck is dog tracking while being driven down the road. Upon inspection, the technician diagnoses the vehicle and notes that the frame is in a diamond condition. Technician A says that towing another truck with a chain attached to one corner of the frame could be the cause. Technician B says the shifting of the front or rear axles could be the cause. Who is correct?

    A. A only
    B. B only
    C. Both A and B
    D. Neither A nor B

50. The LEAST LIKELY purpose for pre-lubing the spindle and hub cavity during hub installation is:

    A. To pre-charge the wheel bearing with lubricant.
    B. To prevent fretting corrosion.
    C. To prevent spindle contamination.
    D. To aid in hub installation.

PREPARATION EXAM 3

1. Which of these statements is correct when diagnosing truck frame problems?

   A. Tandem-axle tractors have the maximum bending moment occur at the bogie centerline.

   B. Single-axle trucks with van bodies have the maximum bending moment occur just ahead of the rear axle.

   C. Frame buckle may be caused by too many holes drilled in the frame web.

   D. Frame twist may be caused by uneven loading.

2. The driver of a truck complains of a shimmy in the front end. A technician performed a front-end alignment. While taking the unit for a road test, the technician finds that a truck still has a shimmy. Technician A says this could be caused by too much positive caster. Technician B says an improper camber setting can be the cause. Who is correct?

   A. A only

   B. B only

   C. Both A and B

   D. Neither A nor B

3. When adjusting caster, all of the following statements are true EXCEPT:

   A. Use only one shim on each side.

   B. The width of the shim should equal the width of the spring.

   C. Do not use shims of more than 1 degree difference from side to side.

   D. Reverse the shims from left to right to correct for a twisted axle.

4. What is one of the standard SAE kingpin sizes used on trailers?

   A. 1.5 inches

   B. 2.5 inches

   C. 3.5 inches

   D. 4.5 inches

5. A frame reinforcement is needed on a tractor. Which of the following is true when using C-channel stock for the repair?

   A. It should be welded directly to the frame rail.

   B. It must be of the same dimensions as the original frame rail.

   C. It must have the same yield strength as the original frame.

   D. It should only be installed on the outside of the original frame rail.

6. While performing an inspection on a power steering system pump, a growling noise is present and air has been found in the system. Technician A says that if the growling noise is still present after the fluid level is checked and the air is bled from the system, the pump bearings or other components may be damaged. Technician B states that when the power steering pump pressure is lower than specified, pump replacement or repair is indicated. Who is correct?

   A. A only
   B. B only
   C. Both A and B
   D. Neither A nor B

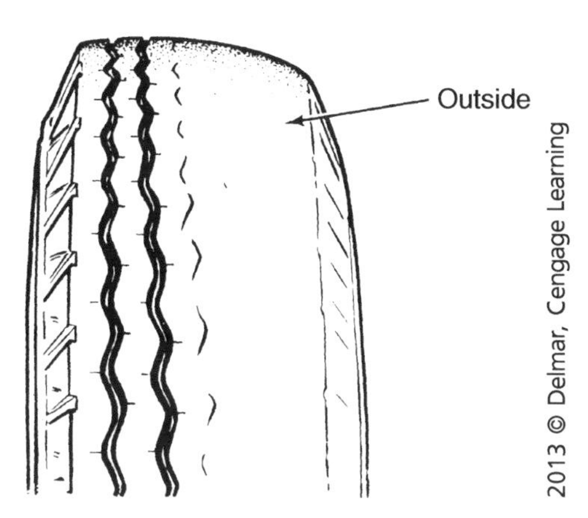

7. Referring to the figure above, what does this tire wear pattern indicate?

   A. Excessive toe-out
   B. Excessive positive camber
   C. Excessive positive caster
   D. Overinflation

8. While performing a front wheel alignment, Technician A says that a camber setting of 1 3/4 degree for the curb side and 1 1/4 degree for the driver side is typical. Technician B says that camber settings depend on the amount of KPI that is built into the axle. Who is correct?

   A. A only
   B. B only
   C. Both A and B
   D. Neither A nor B

9. Technician A says that poppet valves must be adjusted properly to prevent excessive pressure on the steering linkages when the steering wheel is turned fully to the left or right. Technician B says that the poppet valve adjustment is performed with the steering gear installed on the vehicle and the power steering fluid at normal operating temperature. Who is correct?

   A. A only
   B. B only
   C. Both A and B
   D. Neither A nor B

10. Technician A says steer axle misalignment may cause abnormal steer axle tire wear. Technician B says drive axle misalignment may cause abnormal steer axle tire wear. Who is correct?

    A. A only
    B. B only
    C. Both A and B
    D. Neither A nor B

11. A driver complains that his tractor/trailer is leaning to the left and air is heard leaking out of the left-rear height control valve. He also states that the air system pressure went to 0 psi. The cause of the low air pressure in the braking system could be:

    A. A defective pressure protection valve.
    B. A defective pressure reduction valve.
    C. A defective height control valve.
    D. An improper ride height adjustment.

12. When a technician is performing a close visual inspection of the power steering belt(s) his inspection may reveal all of the following EXCEPT:

    A. Proper belt tension.
    B. Premature wear due to misalignment.
    C. Correct orientation of dual belt application.
    D. Proper belt seating in the pulley.

13. A driver complains that his heavy-duty truck requires excessive steering effort and that the steering wheel return is too fast. Which of these is the most likely cause?

    A. Incorrect turning angle
    B. Too much negative camber
    C. Too much positive caster
    D. Underinflated tires

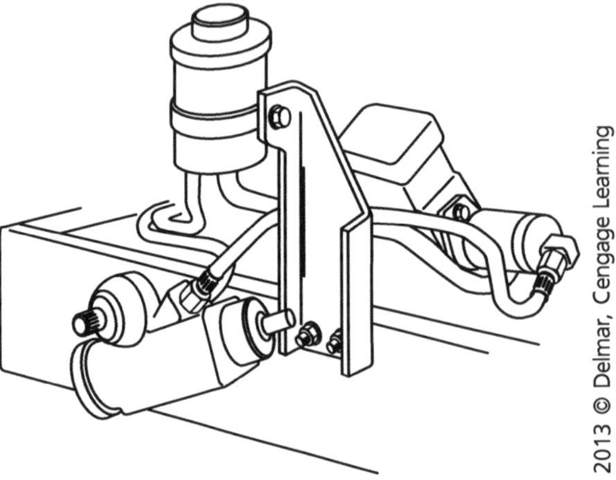

2013 © Delmar, Cengage Learning

14. Referring to the figure above, the power steering pump is being removed. All of the tasks listed below are required to be performed during removal and replacement EXCEPT:

    A. Check the pump mounting holes for wear.
    B. Remove the hoses from the pump and cap the fittings.
    C. Reuse the o-ring if it is in good condition.
    D. Bleed.

15. Toe may be defined as:
    A. The forward or rearward tilt of the kingpin at the top.
    B. The tracking angle of the tires from a true straight-ahead track.
    C. The inward or outward tilt of the top of the wheel when viewed from the front of the vehicle.
    D. The amount in degrees that the top of the kingpin inclines away from vertical as viewed from the front of the vehicle.

16. When installing cups in a cast-iron hub, the hub needs to be heated to between:
    A. 200°F and 220°F.
    B. 450°F and 500°F.
    C. 180°F and 200°F.
    D. Cast-iron hubs are not heated.

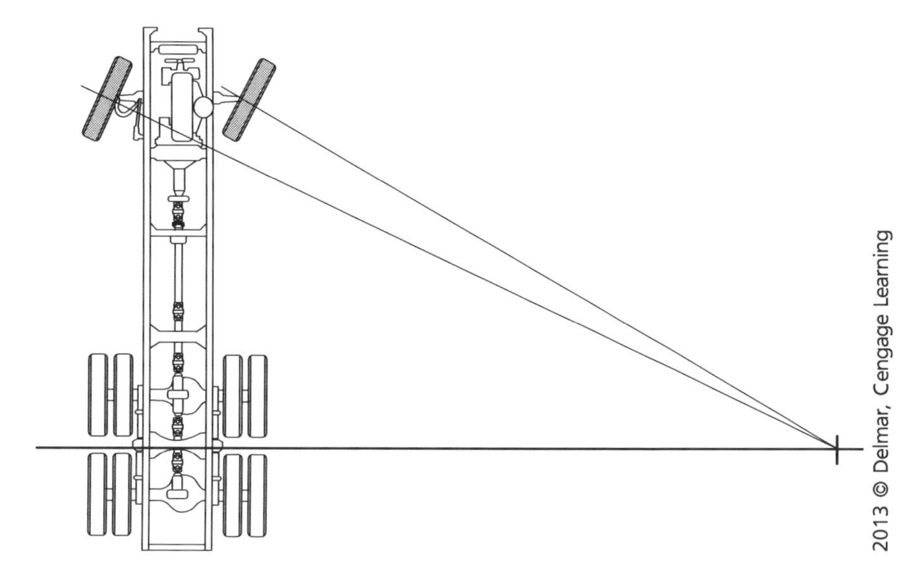

17. Referring to the figure above, Technician A says that most steering knuckles use adjustable wheel stops for setting toe-out on turns. Technician B says that during a turn the front wheels must be turned at different angles to prevent tire scuffing. Who is correct?
    A. A only
    B. B only
    C. Both A and B
    D. Neither A nor B

18. The drag link connects the pitman arm to which component on the front non-drive steer axle?
    A. The steering gear
    B. The steering arm
    C. The front axle
    D. The torque arm

19. Upon inspection, a technician finds broken leaves within the spring pack on a leaf spring-type walking beam suspension. Technician A says that this may be caused by overloading the suspension. Technician B says that the spring pack should be replaced. Who is correct?

    A. A only
    B. B only
    C. Both A and B
    D. Neither A nor B

20. Tire matching is being discussed. Technician A says that dual tires are matched to prevent tire tread wear from slippage from uneven surface areas. Technician B says dual drive wheels may be measured using a tire tape measure. Who is correct?

    A. A only
    B. B only
    C. Both A and B
    D. Neither A nor B

21. Technician A says air spring suspensions use shock absorbers to help maintain axle alignment. Technician B says that shock absorbers are used to control spring oscillations. Who is correct?

    A. A only
    B. B only
    C. Both A and B
    D. Neither A nor B

22. Caster may be defined as:

    A. The forward or rearward tilt of the kingpin at the top.
    B. The tracking angle of the tires from a true straight-ahead track.
    C. The inward or outward tilt of the top of the wheel when viewed from the front of the vehicle.
    D. The amount in degrees that the top of the kingpin inclines away from vertical as viewed from the front of the vehicle.

23. When draining a power steering system, all of the following should be done EXCEPT:

    A. Drain the fluid by removing the return hose at the remote reservoir fitting.
    B. With the engine stopped, turn the steering wheel fully in each direction.
    C. When the clean fluid begins to discharge from the return hose, shut the engine off.
    D. Perform a final "bleed" on the system once the return hose is reinstalled.

24. A technician finds an irregular wear pattern on a tire that is overinflated. Which wear pattern did the technician find on the tire?

    A. Outside edges of the tire
    B. Center of the tire
    C. Cupping pattern
    D. Inside edges of the tire

25. When removing a disc wheel from any truck, Technician A says that when a disc wheel needs to be removed from any truck, the right-side wheel will have right-hand threads and the left side left-hand threads. Technician B says to follow good safety practices by wearing safety glasses and not standing in front of a deflating tire. Who is correct?

    A. A only
    B. B only
    C. Both A and B
    D. Neither A nor B

26. Which of the following shop tools would most likely be required to remove old rubber bushings from an equalizing beam-type suspension?

    A. A hammer and chisel
    B. An oxy-acetylene cutting torch
    C. A rosebud
    D. A 50-ton hydraulic press

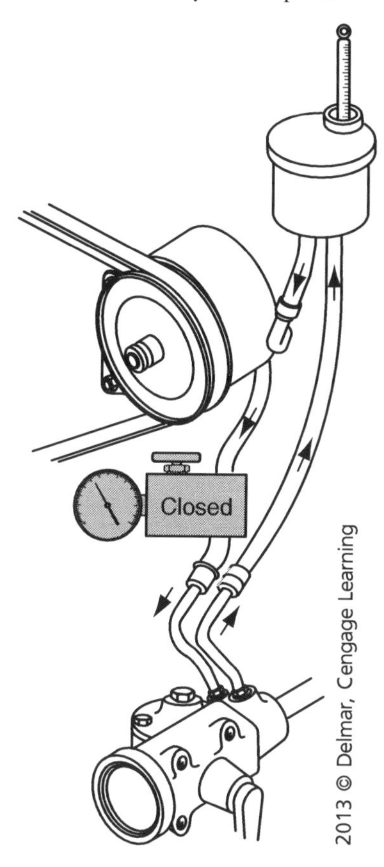

2013 © Delmar, Cengage Learning

27. Referring to the figure above, when using a power steering system analyzer with the engine running, if the flow control valve on the analyzer is closed, what should happen?

    A. The system pressure goes up.
    B. The system pressure goes down.
    C. The wheels turn off-center.
    D. The GPM gauge reading increases.

28. What type of nuts are used to hold the wheels in place on a hub with a hub-piloted mounting system?

    A. A flanged nut
    B. A nut and washer combination
    C. An axle nut
    D. A ball-seat nut

29. You are balancing the radial tires on a medium truck. Technician A says wheel runout should be measured before the balance procedure. Technician B says after the wheels are balanced, mount the tire with the wheel weights 180 degrees from the brake drum weights. Who is right?

    A. A only
    B. B only
    C. Both A and B
    D. Neither A nor B

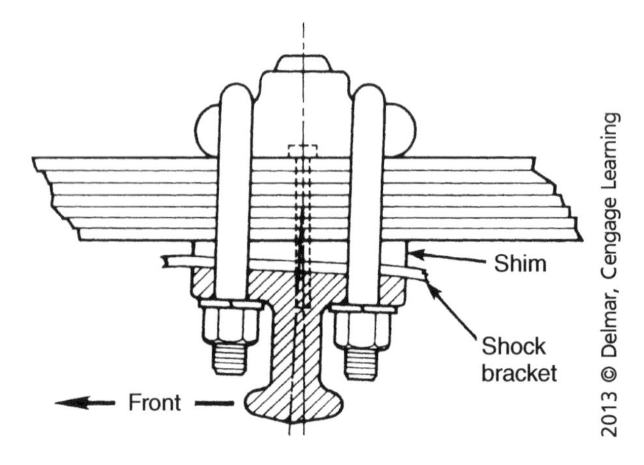

30. Referring to the figure above, a technician has installed the shim to adjust:

    A. Caster.
    B. Camber.
    C. Toe-in.
    D. Spring sag.

31. Two technicians are discussing the replacement procedure for a power steering pump pulley. Technician A states that if there is a retaining nut on the shaft, the pulley is retained by a woodruff key to prevent pulley rotation on the shaft. Technician B says that it is acceptable to use a soft-faced hammer and to tap on the pump shaft to remove the pulley. Who is correct?

    A. A only
    B. B only
    C. Both A and B
    D. Neither A nor B

32. A vehicle is found to be "dog tracking." What would be the LEAST LIKELY cause of this condition?

    A. Loose spring shackles
    B. Loose U-bolts
    C. Bent torque arm
    D. Bent frame

33. Four functions a lubricant provides the wheel bearings are to lubricate, clean the bearings, prevent corrosion, and:

    A. Prevent contamination.
    B. Contain the bearing.
    C. Control the bearing.
    D. Cool the bearing.

34. A technician is inspecting a front axle assembly. He states that a technician should check for a twisted axle beam if any of the following conditions exist EXCEPT:

    A. The difference in caster angle exceeds 1/2 degree from side to side.
    B. The caster shims in place differ by 1 degree or more.
    C. A low-speed shimmy exists and there is no evidence of looseness elsewhere in the steering system.
    D. Excessive tire wear exists.

35. Technician A says that the condition of high hitch occurs when the trailer kingpin and bolster plate are positioned too high in the fifth wheel. Technician B says this condition is caused by improper bolster plate position during the coupling process. Who is right?

    A. A only
    B. B only
    C. Both A and B
    D. Neither A nor B

36. Technician A states that axle alignment on a tandem rear axle suspension is corrected by adjusting the torque arm length. Technician B states that some rear tandem axle suspensions rotate an eccentric adjustment bolt to correct axle alignment. Who is correct?

    A. A only
    B. B only
    C. Both A and B
    D. Neither A nor B

37. A technician turns the steering wheel to the left until the steering effort increases. Technician A states that if the power steering pump goes into the pressure-relief mode, and the clearance between the axle stop and axle stop adjustment screw is incorrect, the poppet valves require adjustment. Technician B states that if the power steering pump goes into the pressure-relief mode, and the clearance between the axle stop and axle stop adjustment screw is incorrect, the axle stops need to be adjusted. Who is correct?

    A. A only
    B. B only
    C. Both A and B
    D. Neither A nor B

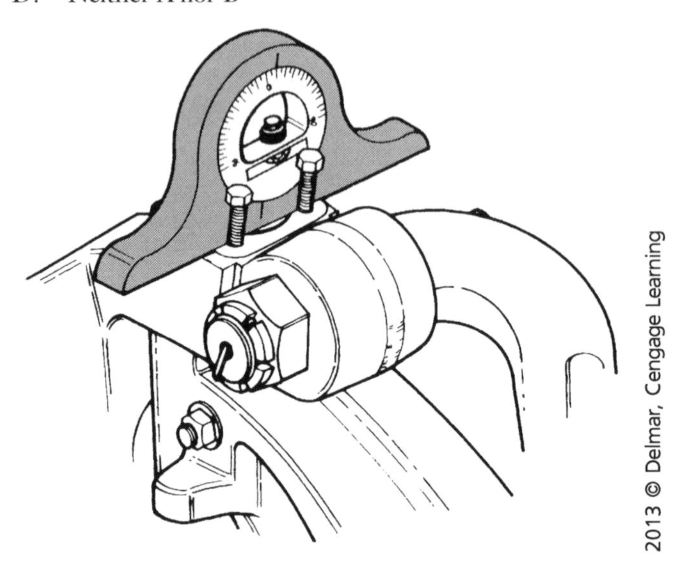

2013 © Delmar, Cengage Learning

38. Referring to the figure above, what alignment angle is being checked?

    A. Camber
    B. Caster
    C. KPI (kingpin inclination)
    D. Steering radius

39. A bumping action is felt combined with erratic steering during braking and acceleration. What would be the LEAST LIKELY cause?

    A. A dry fifth wheel
    B. A worn trailer kingpin
    C. A slack in the fifth wheel
    D. A trailer not properly loaded

40. A driver complains of a front-end shimmy with slight vibrations. Technician A says loose or worn kingpins or kingpin bushings might be the cause. Technician B says that an overloaded trailer might be the cause. Who is correct?

    A. A only
    B. B only
    C. Both A and B
    D. Neither A nor B

41. A truck cab equipped with an air ride suspension leans to one side. Upon inspection, the technician does not find any air leaks. The most likely cause is:

    A. An inoperative air brake control valve.

    B. A blocked cab height control valve intake port.

    C. A kinked air hose to the air bag.

    D. A defective air compressor.

42. A technician is replacing the kingpins on a conventional axle configuration. During the kingpin installation, he encounters interference between the kingpin and the bushing. All of the following could be the cause EXCEPT:

    A. The right-side kingpin was switched with the left-side kingpin.

    B. A burred bushing during installation.

    C. An improperly aligned bushing.

    D. Bushings were not reamed.

43. When installing shock absorbers, if the torque applied to the mounting hardware is inadequate, all of the following may be a result EXCEPT:

    A. Fastener fatigue.

    B. Noisy operation.

    C. Consistent directional stability.

    D. Premature bushing failure.

44. While driving a vehicle at 28 mph, a driver notices a vibration in the steering wheel. Which of the following is the LEAST LIKELY cause?

    A. Improper wheel balance

    B. Worn kingpin bushings

    C. A bent wheel mounting surface

    D. A shifted belt inside a tire

45. A routine inspection shows discoloration of power steering fluid. The LEAST LIKELY cause for this condition might be:

    A. The wrong type of fluid.

    B. Mixed brands of fluid.

    C. Water mixed with fluid.

    D. Overheated or burned smell condition.

46. While replacing the steering knuckle bushings, a technician finds there is interference during the kingpin installation. The LEAST LIKELY cause would be:

    A. A worn axle eye.

    B. A burred bushing during installation.

    C. An improperly aligned bushing.

    D. Omission of the reaming operation.

47. A technician has identified excessive tire wear. Which of the following suspension components is the LEAST LIKELY cause?

    A. A bent spindle
    B. An improper tie rod setting
    C. A worn center cross tube bushing
    D. A preloaded axle wheel bearing

48. What is the LEAST LIKELY cause for a driver to complain about rough ride characteristics?

    A. Excessive positive caster
    B. A leaking or damaged shock absorber
    C. Loose/worn suspension component
    D. Defective rear wheel bearings

49. The LEAST LIKELY cause for a power steering pump pulley to become misaligned is:

    A. An over-pressed pulley.
    B. A loose fit from the pulley hub to pump shaft.
    C. A worn or loose pump mounting bracket.
    D. A broken engine mount.

50. When adjusting caster, what is the LEAST LIKELY procedure to perform?

    A. Use only one shim on each side.
    B. The width of the shim should equal the width of the spring.
    C. Install the tapered shim with the thickest section to the rear of the vehicle.
    D. Reverse one shim to correct for a twisted axle.

PREPARATION EXAM 4

1. Technician A says that if the cab air bag has a leak, the pressure protection valve closes to protect the air brake system from air loss. Technician B says the pressure protection valve protects the air bag if the air brake system has excessive pressure. Who is correct?

   A. A only
   B. B only
   C. Both A and B
   D. Neither A nor B

2. When performing a rear-axle alignment on a vehicle equipped with a tandem axle, Technician A says that the front axle should be aligned to the frame first. Technician B says that the rear axle should be aligned first. Who is correct?

   A. A only
   B. B only
   C. Both A and B
   D. Neither A nor B

3. Center shaft wear may be caused by all of the following EXCEPT:

   A. Beam end bushing wear.
   B. Beam center bushing wear.
   C. Torque arm bushing wear.
   D. Front spring bushing wear.

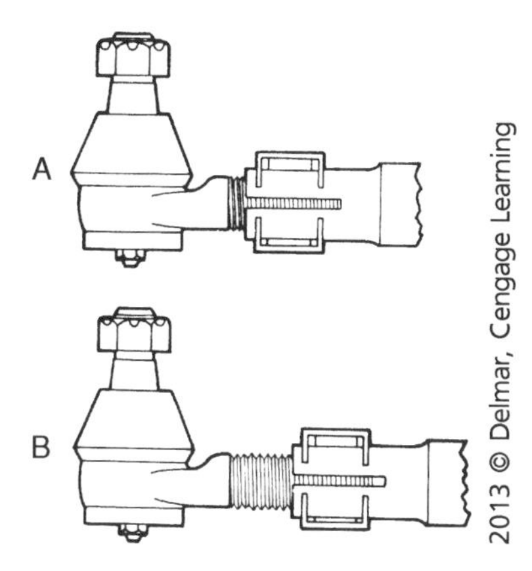

4. Referring to the figure above, you are installing a tie rod end. Which position is the correct installation?

   A. Position A only
   B. Position B only
   C. Both A and B are acceptable
   D. Neither A nor B is acceptable

5. After the front springs were replaced on his tractor, the driver complains that the steering wheel does not return to center after a turn but steers fine straight ahead. The tires show no abnormal wear pattern. The first thing to check should be:

   A. Wheel balance.

   B. Camber adjustment.

   C. Toe-out on turns.

   D. Caster adjustment.

6. When replacing broken U-bolts, which of the following is the LEAST LIKELY procedure to perform?

   A. Check the condition of the leaf spring center bolt.

   B. Torque the U-bolt 10 percent above specs to allow for seating and stretch.

   C. Clean mounting surfaces and lubricate the U-bolt threads.

   D. Re-torque the U-bolts after 1,000 miles of operation.

7. A driver complains of reduced power steering assist. Technician A states that the first thing that should be checked is the power steering fluid level. Technician B says the air in the system could be the cause. Who is correct?

   A. A only

   B. B only

   C. Both A and B

   D. Neither A nor B

8. A technician is dismounting a split ring tire. When inspecting the components, what is the LEAST LIKELY thing to check?

   A. Excessive rust or corrosion buildup

   B. Tire brand

   C. Bent flanges

   D. Deep tool marks on the rings and gutter area

9. A technician notices grease leaking out through the pivot bearing while greasing a front-axle kingpin. What should the technician do next?

   A. Replace the bearing.

   B. Reduce grease injection pressure.

   C. Continue greasing until four drips of grease fall.

   D. Do nothing; it shows thorough distribution of the grease.

10. Excessive camber angles may cause all of the following EXCEPT:

   A. Abnormal tire wear on the inside of the tread.

   B. Abnormal tire wear on the outside of the tread.

   C. Steering instability.

   D. Hard steering going into a turn.

11. The drive tires of a vehicle were replaced at time of PM. After only 10,000 miles, they are worn down to the minimum wear indicators. Which of the following could be the cause?

    A. Damaged/worn rear axle bearings

    B. Rear caster measurement

    C. Front wheel toe-in out of specification

    D. Bent rear axle housing

12. When adjusting the toe setting on a class 8 tractor, increasing the length of a tie rod has what effect to the toe setting?

    A. Toe-out increases

    B. Toe-in decreases

    C. Toe-in increases

    D. Toe remains neutral

13. Technician A says that when inspecting the drive belts on a power steering system, the pulleys should be checked for alignment and wear. Technician B states that when inspecting a power steering system belt, the tension should be relieved and the belt tensioner should be checked for proper operation. Who is correct?

    A. A only

    B. B only

    C. Both A and B

    D. Neither A nor B

14. A driver complains of repeated air spring ruptures on a trailer equipped with a lift axle air suspension. Technician A says that the trailer may be continually overloaded. Technician B says that the maximum air pressure adjustment on the hand control valve may be too high. Who is correct?

    A. A only

    B. B only

    C. Both A and B

    D. Neither A nor B

15. Technician A says that the objective of a toe specification in truck steering systems is to achieve zero toe when the truck is fully loaded and running at highway speeds. Technician B says that before setting toe the camber angle should be adjusted. Who is correct?

    A. A only

    B. B only

    C. Both A and B

    D. Neither A nor B

16. Misaligned drive belt pulleys may cause all of the following EXCEPT:

   A. Low pump pressure.
   B. Wear on the belt edges.
   C. Pulley wear.
   D. Belt squeal.

17. A driver complains of his vehicle leaning in the front end. Upon inspection it is determined that one of the front springs is weak. What is the recommended service procedure?

   A. Align the vehicle using the present ride height.
   B. Add an auxiliary spring for added support.
   C. Replace the sagged spring only.
   D. Replace both springs on the axle as a pair.

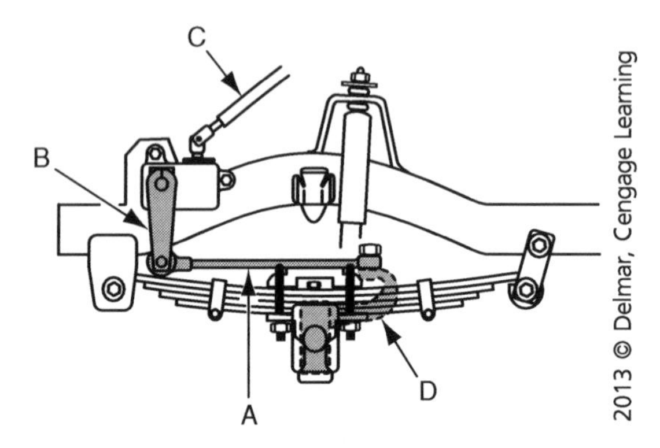

18. Referring to the figure above, the pitman arm is:

   A. Letter A.
   B. Letter B.
   C. Letter C.
   D. Letter D.

19. What is the LEAST LIKELY effect from an excessive camber angle?

   A. Abnormal tire wear on the inside of the tread
   B. Abnormal tire wear on the outside of the tread
   C. Steering instability
   D. Hard steering going into a turn

20. A tractor was involved in a front-end collision and the technician suspects that there may be damage to the front axle. To determine if axle damage exists, the technician should lift both front wheels and rotate the tires to:

   A. Measure for radial runout.
   B. Measure front wheel setback.
   C. Measure relay rod length and compare to axle.
   D. Measure toe-in.

21. The LEAST LIKELY cause of abnormal tire wear, shimmy, or vibration is:
    A. Tire/wheel imbalance.
    B. Excessive wheel or hub runout.
    C. Excessive cam brake stroke.
    D. Improper tire mounting.

22. Technician A says that lift axles are always a fixed straight axle design. Technician B says that lift axles are designed to be either a fixed axle design or a steerable axle design. Who is correct?
    A. A only
    B. B only
    C. Both A and B
    D. Neither A nor B

23. When inspecting a power steering gear before removal, which of these tasks would be the LEAST LIKELY task to perform?
    A. Rotate the input shaft and visually determine if it is true.
    B. Clean and inspect for evidence of fluid leakage.
    C. Check all mounting fasteners.
    D. Adjust the truck toe.

24. What is the torque specification for the spindle nut on a PreSet hub?
    A. 50 ft-lbs
    B. 200 ft-lbs
    C. 300 ft-lbs
    D. 750 ft-lbs

25. While adjusting front wheel caster:
    A. Excessive positive caster decreases steering effort.
    B. Excessive positive caster may cause front-wheel shimmy.
    C. The front suspension caster becomes more negative when the rear suspension height is lowered.
    D. Excessive negative caster results in harsh ride characteristics.

26. A truck frame is being inspected while on a frame-straightening rack. Technician A says buckle is a condition where one side rail is bent upward from its original position. Technician B says that sag occurs when the frame or one side rail is bent downward from its original position. Who is correct?
    A. A only
    B. B only
    C. Both A and B
    D. Neither A nor B

27. When airing tires being restrained in a device, all of the following should be used EXCEPT:
    - A. A clip-on chuck.
    - B. An in-line valve with pressure gauge or pre-settable regulator.
    - C. Enough hose between the clip-on chuck and the in-line valve to allow the user to stand outside of the trajectory.
    - D. A handheld pressure gauge.

28. The LEAST LIKELY cause of a front steer axle pull condition is:
    - A. A dragging brake.
    - B. An out-of-adjustment brake.
    - C. Incorrect brake timing.
    - D. An incorrect crack pressure relay valve.

29. A driver complains of a rough ride and body sway on his tractor. While performing an inspection, a faulty shock absorber is found. Technician A says replace the shock absorbers in pairs. Technician B says if the shock absorbers are not leaking they are still good. Who is correct?
    - A. A only
    - B. B only
    - C. Both A and B
    - D. Neither A nor B

30. After performing a front-axle and linkage binding test, the technician notes that the wheel and tire do not return to the straight-ahead position. Which of these components should the technician check?
    - A. Kingpin bearings
    - B. Front suspension springs
    - C. Rear suspension springs
    - D. Shock absorbers

31. While checking the fluid level in a power steering system, Technician A says that foaming in the remote reservoir may indicate air in the system. Technician B says that most original equipment manufacturers (OEMs) recommend that the fluid be warmed up prior to checking the level. Who is correct?
    - A. A only
    - B. B only
    - C. Both A and B
    - D. Neither A nor B

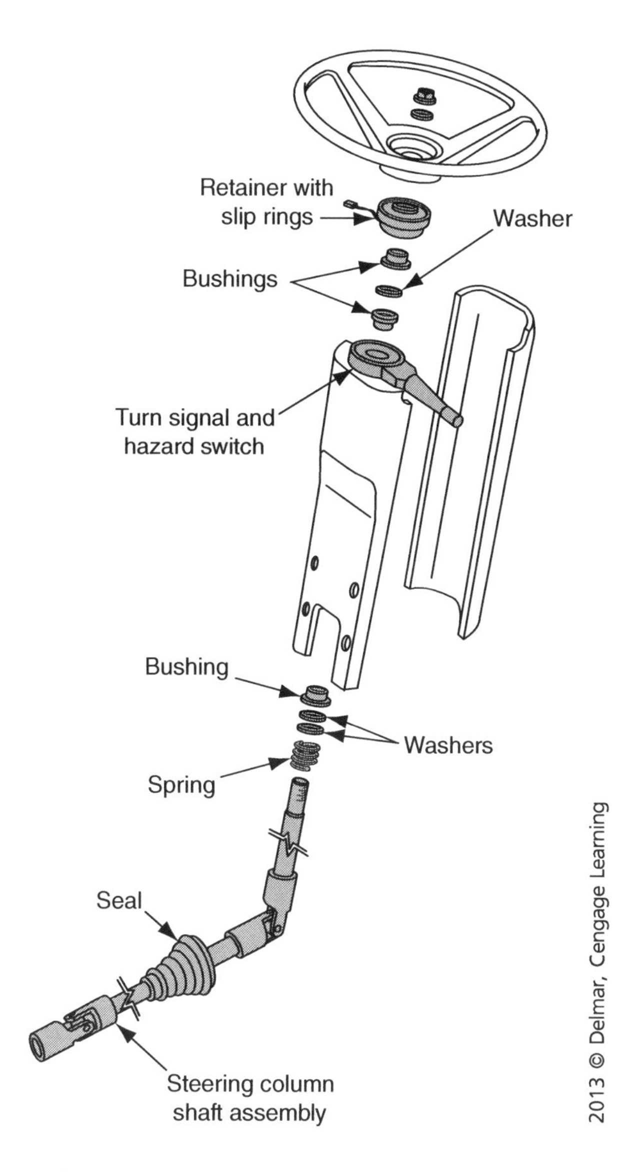

32. Referring to the figure above, noise in the steering column is being discussed. Technician A says that noise might be caused by the upper or lower shaft bearing being tight or frozen. Technician B says that noise might be caused from a steering shaft snap ring not being seated. Who is correct?

    A. A only
    B. B only
    C. Both A and B
    D. Neither A nor B

33. While turning left or right on a vehicle equipped with power steering, the driver hears a squeaking noise. This might be caused by:

    A. A dry kingpin pivot bearing.
    B. A bent front wheel.
    C. A fluid reservoir overfilled.
    D. Worn front shock absorbers.

34. After mounting a rim on a spoke wheel on a drive axle, a technician checks the lateral runout and finds it exceeding the maximum allowable runout specification of 0.125 inches. The first step to correct this procedure should be:

    A. Remove the wheel, demount the tire, and straighten the wheel in a hydraulic press.
    B. Loosen the stud nut at the point of greatest clearance, and tighten the stud nut opposite this nut.
    C. Loosen and re-torque all the stud nuts to specifications.
    D. Clean, inspect, lubricate, and adjust the wheel bearings.

35. When installing a hub cap, all of the following requirements need to be followed EXCEPT:

    A. Use SAE Grade 5 bolts or stronger.
    B. Do not use star washers.
    C. Do not use split lock washers.
    D. Pack the hub cap with grease.

36. Two technicians are discussing front wheel alignment angles. Technician A says the included angle is the sum of the KPI angle and the positive camber angle. Technician B says if the positive camber angle is increased the included angle is decreased. Who is correct?

    A. A only
    B. B only
    C. Both A and B
    D. Neither A nor B

37. An axle alignment on a vehicle equipped with a tandem axle with a walking beam suspension is being performed. Technician A says that if there is axial movement at the center cross shaft, the beam bushings are worn and need to be replaced prior to the alignment being performed. Technician B says that the rear axle should be aligned to the frame first, and then the front axle aligned to the rear axle. Who is correct?

    A. A only
    B. B only
    C. Both A and B
    D. Neither A nor B

38. Oil leaks may occur in a power steering gear in all of the locations listed EXCEPT:

    A. Side cover o-ring.
    B. Pitman shaft oil seal.
    C. Top cover seal.
    D. Reservoir o-ring.

39. While performing a routine PM inspection, a technician notes that there are cupping marks around the circumference of the tire. The most likely cause of this tire condition is:

    A. Shock absorbers.
    B. Camber.
    C. Toe-out on turns.
    D. Toe-in.

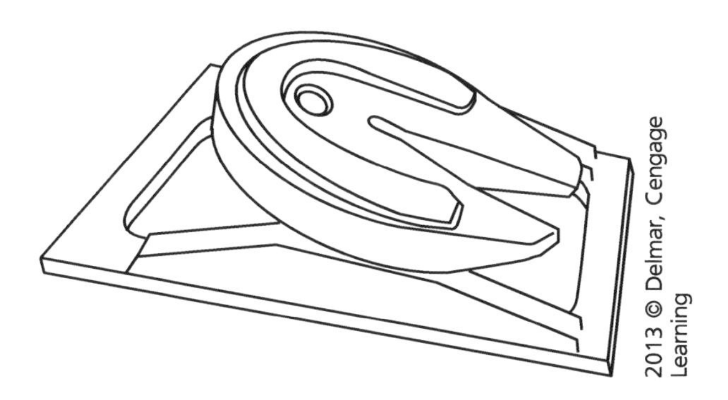
2013 © Delmar, Cengage Learning

40. Referring to the figure above, a technician is lubricating the fifth wheel. What is the recommended type of lube he should use?

    A. Water-resistant lithium-based grease
    B. Sodium soap-based grease.
    C. Calcium soap-based grease
    D. Non-petroleum lubricant

41. Technician A says that a bowed upper coupler is caused by trailer overloading. Technician B says that a bowed upper coupler might be the result of using a light gauge steel material. Who is correct?

    A. A only
    B. B only
    C. Both A and B
    D. Neither A nor B

42. Technician A says that if the rear suspension height is lowered, the front wheel caster becomes more negative. Technician B says that if the camber is negative, the camber angle must be added to the KPI angle to obtain the included angle. Who is correct?

    A. A only
    B. B only
    C. Both A and B
    D. Neither A nor B

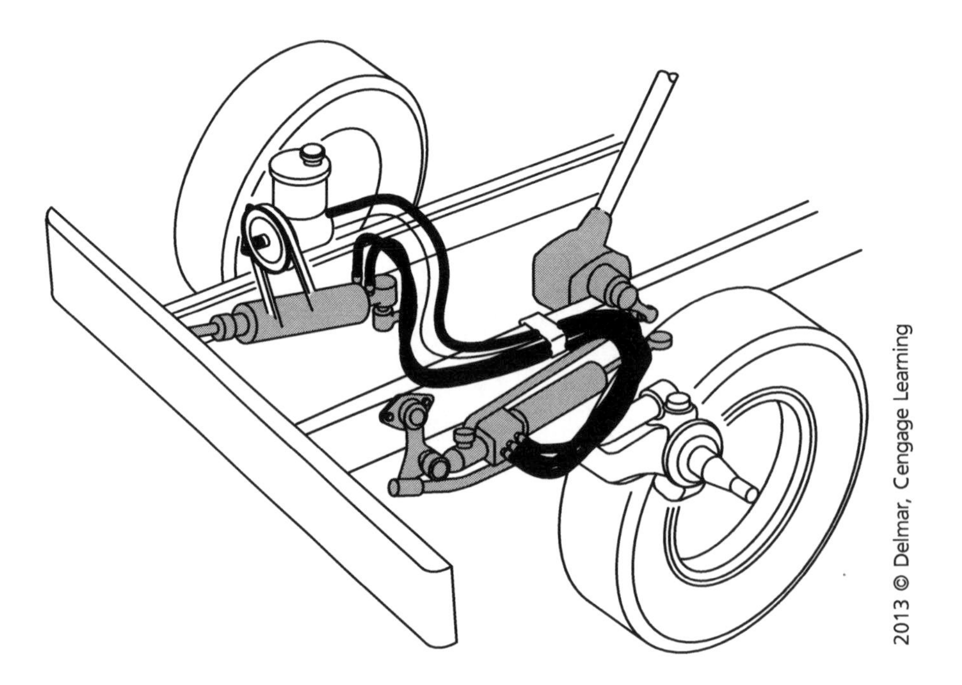

43. The linkage-assist-type power steering shown in the above figure binds when turning corners in either direction, but with short steering corrections, wheel recovery is normal. Which of these is the LEAST LIKELY cause?

A. Worn kingpins
B. Improper sector lash adjustment
C. A bent worm gear
D. Worn-out tie rod end assembly

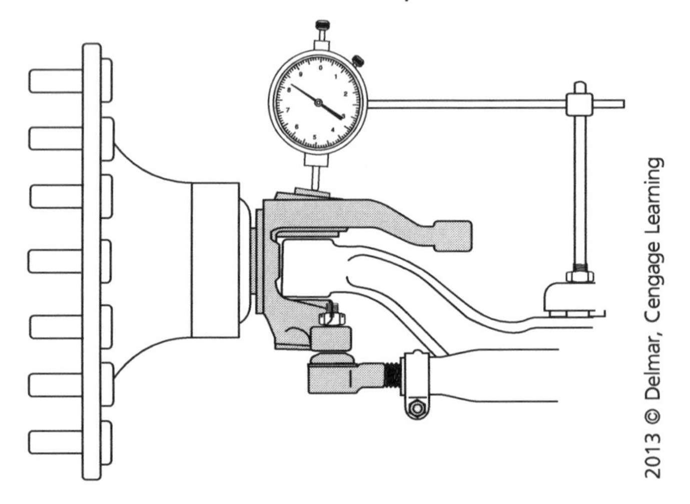

44. With the dial indicator set up as shown in the above figure, what is being checked?

A. Camber
B. Ball joint end-play
C. Steering knuckle vertical play
D. Lower bearing free-play

45. When the height control lever of the suspension height control valve (leveling valve) is raised upward off of horizontal on an air ride suspension, what happens to the air pressure within the air bags?

    A. The air is exhausted.
    B. The bag is further inflated.
    C. The suspension air circuit is dumped.
    D. The chassis system pressure is reduced.

46. When discussing end-play on a PreSet hub, Technician A says when the end-play is beyond 0.006 inches the hub requires servicing. Technician B says that a PreSet hub is not serviceable and needs to be replaced. Who is correct?

    A. A only
    B. B only
    C. Both A and B
    D. Neither A nor B

47. The LEAST LIKELY cause of hard steering is:

    A. A dry fifth wheel.
    B. An overloaded steer axle.
    C. A contaminated power steering system.
    D. Leaking shock absorbers.

48. The steering wheel fails to return to the top tilt position. All of the following might cause the problem EXCEPT:

    A. Bound up pivot pins.
    B. A faulty tilt spring.
    C. Turn signal wires that are too tight.
    D. Worn tilt bumpers.

49. All of the following are methods of adjusting torque arm length while performing an axle alignment EXCEPT:

    A. Shims between the torque arm and front hanger.
    B. Rotating the threaded portion of the torque arm.
    C. Rotating an eccentric bolt to vary the length.
    D. Replace the torque arm with a longer one.

50. A truck tire is found to have been run in a very underinflated condition. Technician A says that the tire may be inflated without removing it from the truck. Technician B says that before balancing a tire, wheel runout is checked. Who is correct?

    A. A only
    B. B only
    C. Both A and B
    D. Neither A nor B

PREPARATION EXAM 5

1. All of the following tire wear conditions are considered abnormal EXCEPT:
   A. Tread river erosion.
   B. Excessive toe-in wear.
   C. Excessive toe-out wear.
   D. Excessive shoulder wear.

2. The LEAST LIKELY cause of premature suspension bushing wear is:
   A. A broken leaf spring center bolt.
   B. Excessive rear axle castor.
   C. Excessive wear to adjacent suspension components.
   D. Loose U-bolts.

3. While adjusting front wheel camber:
   A. If the front wheel has a positive camber angle, the camber line is tilted inward from the true vertical centerline of the wheel and tire.
   B. Excessive positive camber on the front wheel causes premature tire wear on the inside tread of the tire.
   C. Excessive negative camber on a front wheel causes premature wear on the outside of the tire tread.
   D. Improper camber angle on an I-beam front suspension may be caused by a bent axle or spindle.

4. A vehicle with power steering has a high turning effort at idle. At any other RPM, it operates normally. What is the LEAST LIKELY cause of this condition?
   A. A sticking rotary valve
   B. Excessive wear between the vanes and rotor of the power steering pump assembly
   C. A sticking flow control valve
   D. An obstruction or kink in the fluid return line

5. While diagnosing KPI and front spindle movement:
   A. When the steering wheel is turned, the front spindle movement is parallel to the road surface.
   B. The KPI angle has no effect on steering wheel returning force.
   C. The KPI angle tends to maintain the wheel in a straight-ahead position.
   D. An increase in the KPI angle decreases steering effort.

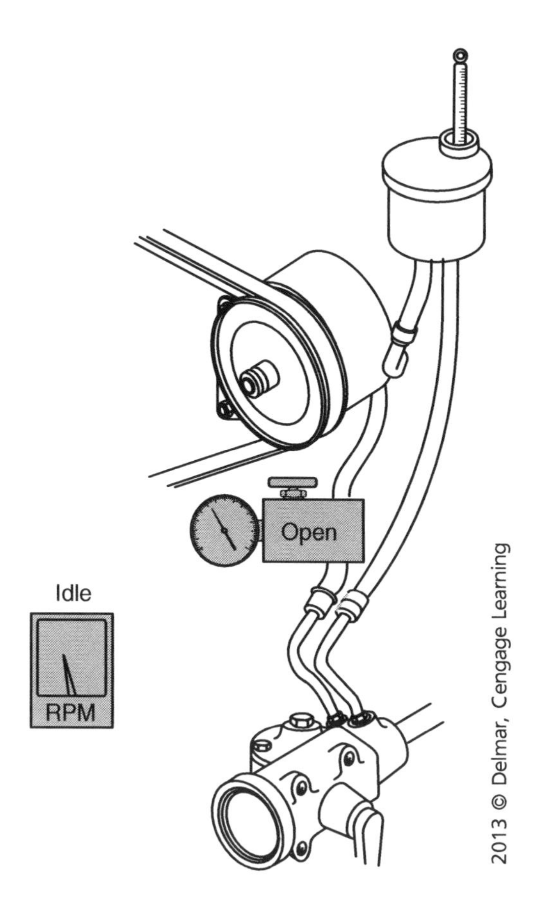

6. Referring to the figure above, two technicians are discussing the procedure for diagnosing a steering system using a steering system analyzer. Technician A says that when the engine is at idle and the gate valve on the analyzer is open, if the pressure is less than 200 psi there may be a restriction in the high-pressure hose. Technician B says that a reading below 200 psi at idle may indicate that the power steering gear is damaged. Who is correct?

   A. A only
   B. B only
   C. Both A and B
   D. Neither A nor B

7. An equalizing-beam suspension may also be referred to as:

   A. A torsion bar.
   B. A walking beam.
   C. An air spring.
   D. A multi-leaf variable rate.

8. A driver complains that his tractor bottoms out when hitting a bump. The LEAST LIKELY cause would be:

   A. Excessive weight on the vehicle.
   B. One or more broken leaves in the spring pack.
   C. Weak or fatigued spring assembly.
   D. Broken or missing rebound clips.

9. What is the established industry torque specification for wheel nuts?

   A. 400–450 ft-lbs
   B. 350–400 ft-lbs
   C. 450–500 ft-lbs
   D. 550–600 ft-lbs

10. While discussing turning radius:

    A. The turning radius is affected by the length of the tie rod.
    B. The turning radius is determined by the steering arm design.
    C. Improper turning radius has no effect on tire tread wear.
    D. During a turn, the inside tire is actually behind the outside tire.

11. Technician A says noise from a manual steering gear might be caused by misalignment of the steering column input shaft. Technician B says that noise coming from the manual steering gear assembly of a linkage-assist-type power steering when turning the steering wheel may be caused by low lubricant level. Who is correct?

    A. A only
    B. B only
    C. Both A and B
    D. Neither A nor B

12. When using the dial indicator for checking the upper and lower kingpin bushing wear on both conventional and unitized wheel end axles, if the upper bushing is worn or damaged what do you replace?

    A. One bushing in the knuckle
    B. The entire axle
    C. Both bushings in the knuckle
    D. The steering knuckle

13. All of the statements regarding front-wheel toe are true EXCEPT:

    A. Driving forces tend to move the front wheels toward a toe-out position on an I-beam front suspension.
    B. Improper front-wheel toe causes featheredge tread wear on the front tires.
    C. Adjusting the front wheels on an I-beam front suspension to a toe-in position improves directional stability.
    D. Front-wheel toe setting on an I-beam front suspension does not affect steering effort.

14. After measuring kingpin inclination (KPI) on a tractor with an I-beam front axle, the technician finds the KPI on the left side is more than specified. The cause of the problem might be:

    A. The knuckle pin may be loose in the end of the axle.
    B. The axle may be bent upward in the center of the axle.
    C. The steering arm may be loose in the steering knuckle.
    D. The drag link may be the wrong length.

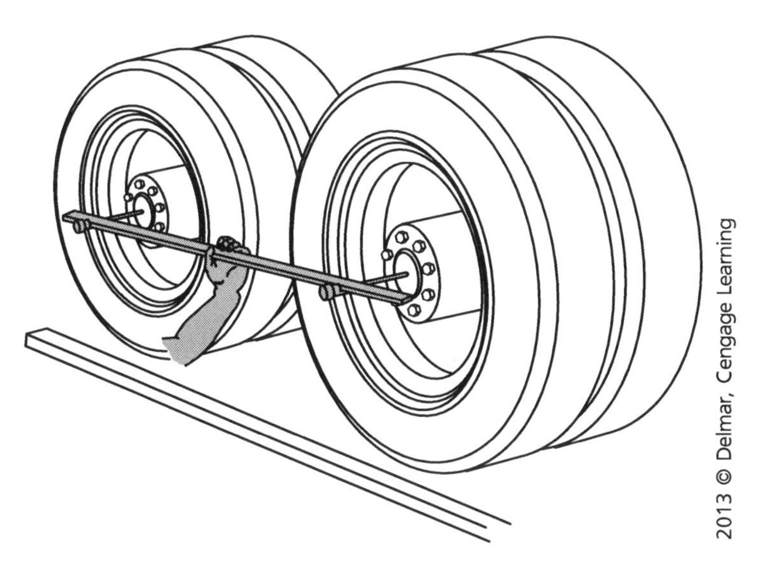

15. Referring to the figure above, the technician is performing an alignment on a tandem-axle vehicle. He finds the distance between the front and rear tandem axle centers is 0.75 inches (19.05 mm) more on the left side compared to the right side. This could be caused by:

    A. A bent rear axle shaft.
    B. Worn rear wheel bearings.
    C. A bent lower torque rod.
    D. An improperly mounted wheel.

16. A frame rail is found to be damaged. Technician A says that in order to reinforce the frame rail, a fishplate frame reinforcement may be used but it should not extend below the frame. Technician B says that a common method for frame reinforcement is to install another section of C-channel, but only on the outside of the frame. Who is correct?

    A. A only
    B. B only
    C. Both A and B
    D. Neither A nor B

17. The LEAST LIKELY result from negative caster is:

    A. Reduced directional stability.
    B. Increased steering wheel returning force.
    C. Reduced steering effort.
    D. Improved ride quality.

18. When discussing air lift axles, all of the following are true statements EXCEPT:

    A. The handle tension is not adjustable on the hand control valve.
    B. The maximum pressure in the air springs is determined by the axle load and model of the suspension.
    C. The minimum pressure setting on the hand control valve determines the air pressure in the air spring with the axle lifted.
    D. The maximum pressure setting on the hand control valve determines the air pressure in the air spring with the axle lifted.

19. A technician found the steering arm bent on the driver's side. After replacement, which repair procedure would be the LEAST LIKELY for the technician to perform?

 A. Perform a road test when repairs are completed.
 B. Replace both outer tie rod ends.
 C. Check and correct changes in wheel alignment.
 D. Lube the replacement part after installation.

20. A tire and wheel assembly is being removed from the vehicle for inspection. Technician A says if you mark the tire to rim position on removal and reinstall the tire in the same position on the rim, the tire will retain dynamic balance. Technician B says that if a vulcanized repair is made to a tire you do not need to perform a dynamic balance. Who is correct?

 A. A only
 B. B only
 C. Both A and B
 D. Neither A nor B

21. The LEAST LIKELY condition that would affect camber is:

 A. Wheel jounce.
 B. Wheel rebound.
 C. Road crown.
 D. A bent Ackerman arm.

22. A hardened truck frame is being discussed. Technician A says gussets may be used when attaching the reinforcement plate to the frame. Technician B says when installing frame reinforcement plates, the original bolt holes in the frame should be used. Who is correct?

 A. A only
 B. B only
 C. Both A and B
 D. Neither A nor B

23. Excessive worm gear end-play may cause:

 A. Lack of lubrication.
 B. A steering fluid leak.
 C. Lost motion within the steering gear.
 D. Hard steering condition during cold operation.

24. Technician A says that when replacing the shaft seal on a power steering pump with an integral reservoir, the pump must be disassembled to facilitate seal installation. Technician B says that when replacing the shaft seal on a power steering pump with a remote reservoir the seal may be serviced without disassembly of the pump. Who is correct?

 A. A only
 B. B only
 C. Both A and B
 D. Neither A nor B

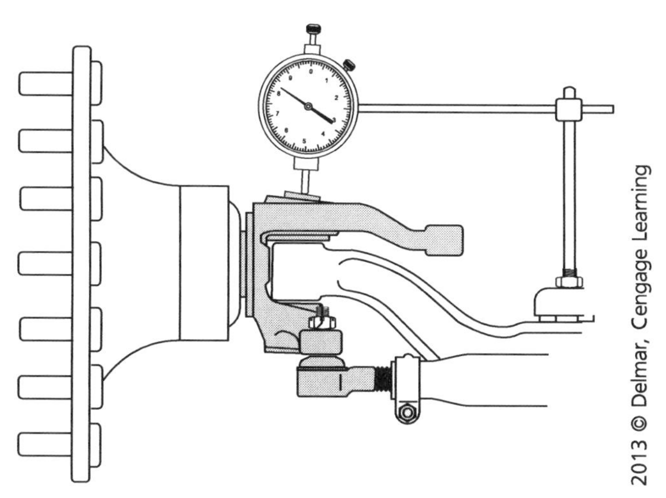
2013 © Delmar, Cengage Learning

25. Referring to the figure above, when performing steering knuckle vertical end-play dial indicator checks for in-service axles, the reading must be between which of the following?

    A. 0.100 and 0.065 inches

    B. 0.010 and 0.065 inches

    C. 0.001 and 0.650 inches

    D. 0.001 and 0.065 inches

26. The rear wheel and hub have been removed as an assembly. The LEAST LIKELY task to be performed when reinstalling the rear wheels and hubs is:

    A. Use a wheel dolly to install the dual wheel assembly.

    B. Use the OEM-recommended lube on the spindle.

    C. Pack the wheel bearings and hub cavity with grease.

    D. Set the wheel bearing end-play after installation.

27. A driver complains of a rattling noise on the right side of a medium-duty truck when driving over bumps. Which of these is the most likely cause?

    A. An overtorqued shock absorber mount

    B. An underinflated right-rear tire

    C. A broken leaf spring rebound clip

    D. A cracked air brake supply line

28. When checking the tie rod boot for cracks, tears, or other damage, also check the retaining nut to ensure which component listed below is installed.

    A. Boot protector

    B. Snap ring

    C. Clevis pin

    D. Cotter pin

29. After rear suspension work was performed, a truck is now experiencing driveline vibrations. Technician A states that loose U-bolts may be causing the vibration. Technician B says that incorrect installation of the axle shims could cause the vibration. Who is correct?

    A. A only
    B. B only
    C. Both A and B
    D. Neither A nor B

30. While discussing wheel alignment variables, Technician A says that road crown is a variable that affects wheel alignment. Technician B says that vehicle loads and unequal weight distribution will affect wheel alignment. Who is correct?

    A. A only
    B. B only
    C. Both A and B
    D. Neither A nor B

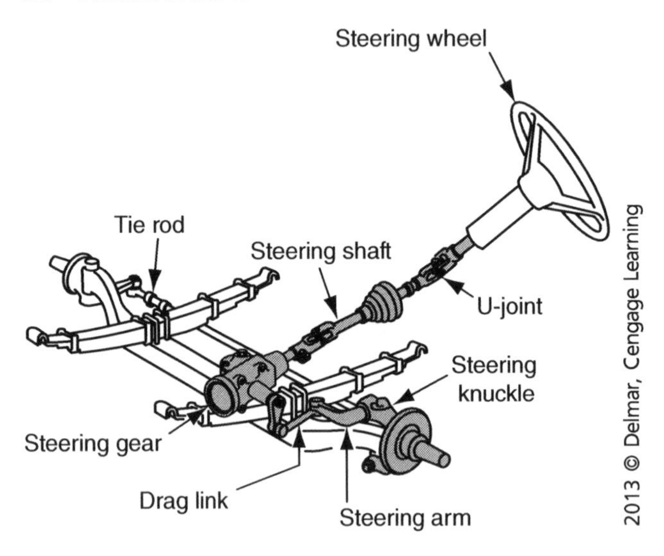

31. Referring to the figure above, two technicians are discussing excessive steering wheel free-play. Technician A says a worn steering shaft universal joint could cause this free-play. Technician B says that the cause could be a bent idler arm. Who is correct?

    A. A only
    B. B only
    C. Both A and B
    D. Neither A nor B

32. During the balancing of a truck tire, a heavy spot is found on the outside edge of the tire. Technician A says when using a static balancing process, a wheel weight is installed 180 degrees from this heavy spot on the outside edge. Technician B says the Technician does not have to enter any information in an electronic wheel balancer. Who is right?

    A. A only
    B. B only
    C. Both A and B
    D. Neither A nor B

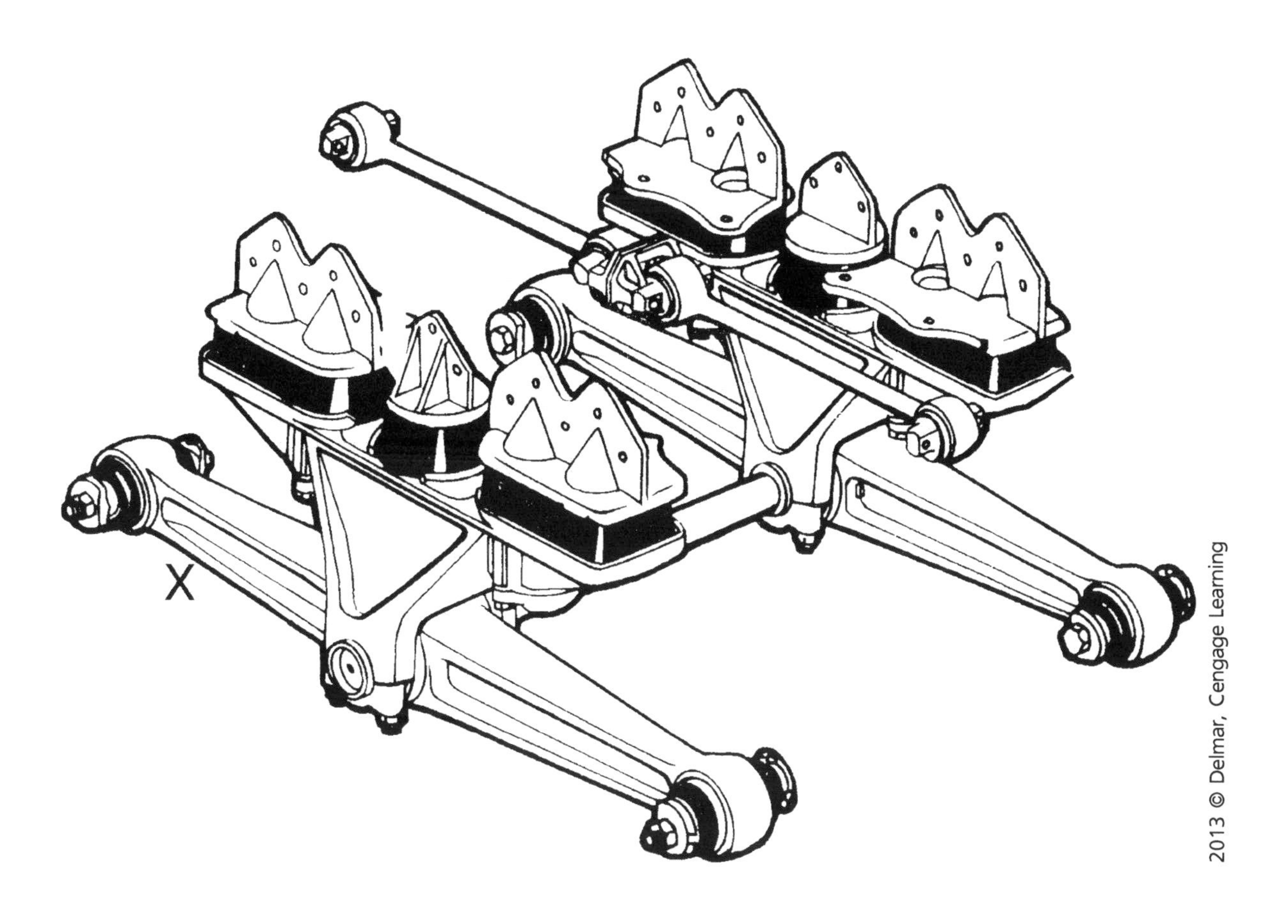

2013 © Delmar, Cengage Learning

33. The figure above shows a rubber load cushion-type equalizing beam suspension. Which of the following best describes how driving, braking, and cornering forces are transmitted to the frame?

    A. Transmitted directly through the rubber cushions to the frame brackets
    B. Transmitted through rubber-bushed vertical drive pins in each cushion
    C. Torque rods transmit 100 percent of these forces
    D. Equalizing beams transmit 100 percent of these forces

34. Two technicians are discussing toe-out on turns. All of the statements are true EXCEPT:

    A. Toe-out on turns is the turning angle of the wheel on the inside of the turn compared with the turning angle of the wheel on the outside of the turn.
    B. When the front wheel on the inside of the turn has turned 18 degrees outward, the front wheel on the outside of the turn may have turned 20 degrees.
    C. Turning radius is the amount of toe-out on turns.
    D. Toe-out on turns is determined by the steering arm design.

35. What is the proper method for checking wheel end-play?

    A. Remove wheels and drum and measure using a dial indicator.
    B. Keep wheels and drum on the vehicle and measure using a dial indicator.
    C. Remove wheels and drum and measure using a tape measure.
    D. Remove wheels and drum and measure by feel.

36. A driver complains that the steering wanders while driving straight down the road. This may be caused by:

A. A tight worm shaft bearing preload.
B. A loose pitman arm shaft over center preload adjustment.
C. A binding universal joint connected to the steering gear.
D. Contaminated power steering fluid.

37. Kingpin inclination (KPI) is being discussed. Technician A says that KPI is the inward tilt of the axle eye in relationship to the true vertical tire centerline as viewed from the front. Technician B says that if the KPI measurement is out of specification on an I-beam front axle, it may be adjusted. Who is correct?

A. A only
B. B only
C. Both A and B
D. Neither A nor B

38. After the first 6,000 miles of new vehicle operation, the draw key nuts should be tightened to which of the following?

A. 15–30 lb-ft
B. 30–45 lb-ft
C. 45–60 lb-ft
D. 60–70 lb-ft

39. Two technicians are discussing split rim tire and wheel assemblies. Technician A says that dismounting and mounting split side rims is extremely dangerous and should only be performed by trained professionals according to OSHA rules and regulations. Technician B says to never hammer on split-side rim rings. Who is correct?

A. A only
B. B only
C. Both A and B
D. Neither A nor B

40. When discussing disk wheels, all of the following statements are true EXCEPT:

A. Disk wheels may have a brake drum that may be removed before the hub and bearings.
B. Disk wheels are mounted on the same bolts as the brake drums.
C. Disk wheels may be retained with wheel nuts having a ball seat.
D. Disk wheels may have retaining bolts that act as pilots to position the wheel.

41. The steering on a tractor pulls to the left while driving straight ahead and there is no indication of excessive tire wear. Technician A says that the left-front wheel may have more positive caster compared to the right-front wheel. Technician B says the left-front wheel may have excessive positive camber. Who is correct?

A. A only
B. B only
C. Both A and B
D. Neither A nor B

42. All of these statements about truck frame defects are true EXCEPT:

    A. A frame sag occurs when the frame rails are bent downward in relation to the rail ends.

    B. A frame bow occurs when one or both frame rails are bent upward in relation to the ends of the rails.

    C. Vehicle tracking is not affected when one frame rail is pushed rearward in relation to the opposite frame rail.

    D. Frame twist occurs when the end of one frame rail is bent upward or downward in relation to the opposite frame rail.

43. A driver complains of noise coming from the steering gear. The LEAST LIKELY cause for steering gear noise is:

    A. A loose pitman shaft cover preload adjustment.

    B. Cut or worn rings on the spool valve.

    C. Loose steering gear mounting bolts.

    D. Tight steering shaft U-joints.

44. Repairs are being performed on a medium-duty truck with a rear air suspension. Technician A says that a trammel bar can be used to measure from a straightedge placed at 90 degrees to the frame rail to the front drive axle. Technician B says if the distance from one side of the axle to a fixed straightedge is different, replace the air spring on that side. Who is correct?

    A. A only

    B. B only

    C. Both A and B

    D. Neither A nor B

45. Two technicians are discussing turning radius. All of these statements are true EXCEPT:

    A. Turning radius is controlled by the Ackerman arms.

    B. You adjust stop bolts to limit it.

    C. It will affect tire tread wear.

    D. The inside wheel is parallel to the outside wheel.

46. Technician A says that when a sliding fifth wheel is moved forward on a tractor, the percentage of total vehicle weight supported by the trailer is reduced. Technician B says that when a tractor sliding fifth wheel is moved rearward, weight is removed from the steer axle. Who is correct?

    A. A only

    B. B only

    C. Both A and B

    D. Neither A nor B

47. Two technicians are discussing wheel balancing. All of the statements are true EXCEPT:
    A. When a wheel and tire have proper static balance, the weight is distributed equally around the axis of wheel and tire rotation.
    B. Improper static wheel balance causes wheel tramp.
    C. When a tire and wheel have proper dynamic wheel balance, the weight of the tire and wheel is distributed equally on both sides of the wheel center.
    D. Improper static wheel balance causes wheel shimmy.

48. A front-end alignment is being performed on a heavy-duty tractor with an I-beam front axle. Technician A checks and corrects the toe setting first before checking and adjusting any other alignment angle. Technician B says that when using a tram bar and tape, you should make the measurement between the tires at both the front and rear of the tire and the measurement should be taken at spindle height. Who is correct?
    A. A only
    B. B only
    C. Both A and B
    D. Neither A nor B

49. While discussing collapsible steering column damage, Technician A says that if the vehicle was involved in a front-end impact, the injected plastic inside of the gearshift tube may be damaged. Technician B says that if the injected plastic is sheared inside the steering shaft, the shaft must be replaced. Who is correct?
    A. A only
    B. B only
    C. Both A and B
    D. Neither A nor B

50. Technician A says that when installing external components to the frame, a technician should use existing bolt holes whenever possible. Technician B says that if new holes must be drilled, avoid drilling holes close to the neutral fiber of the frame rail. Who is correct?
    A. A only
    B. B only
    C. Both A and B
    D. Neither A nor B

PREPARATION EXAM 6

1. What is the LEAST LIKELY condition that improper wheel balance would cause?
   A. Cupped tire tread wear
   B. Vibration while braking
   C. Wheel hop or tramp
   D. Wheel shimmy

2. When checking the end-play measurement on a PreSet hub, the tip of the dial indicator should be placed on the hub cap flange or axle mounting flange, and the magnetic base of the dial indicator should be mounted on the hub's:
   A. Spindle end.
   B. Hub cap.
   C. Fill hole.
   D. Axle housing.

3. An improperly adjusted air suspension ride height can cause:
   A. Damaging driveline angles.
   B. A hissing noise during compressor operation.
   C. Back pressure in the supply line.
   D. A leak, causing intermittent operation.

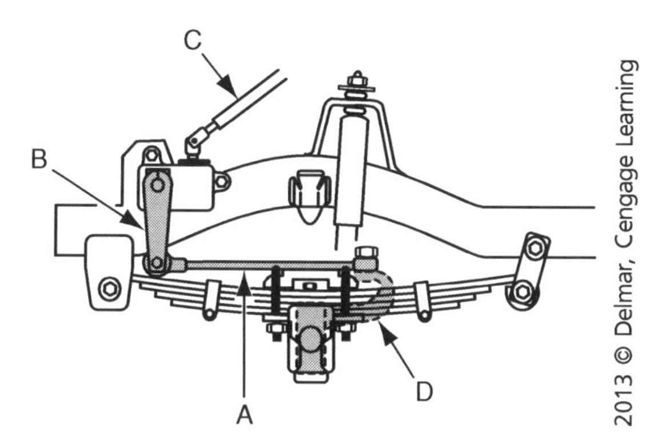

4. Referring to the figure above, the steering arm is:
   A. Letter A.
   B. Letter B.
   C. Letter C.
   D. Letter D.

5. What is the LEAST LIKELY advantage for the use of sliding fifth wheel assemblies on over-the-road equipment?
   A. Weight-over-axle on the tractor may be altered.
   B. Weight-over-axle on the trailer may be altered.
   C. The trailer bridge formula may be altered.
   D. It allows for the use of a rigid-mount fifth wheel.

6. All of the following conditions may cause steering wander or pull EXCEPT:
   A. Improper rear-axle alignment.
   B. Loose steering gear.
   C. Excessive positive caster.
   D. Worn kingpin bushings.

7. When adjusting front-wheel camber:
   A. If the front wheel has a positive camber angle, the camber line is tilted inward from the tire vertical centerline of the wheel and tire.
   B. Excessive positive camber on a front wheel causes premature wear on the inside edge of the tire tread.
   C. The vehicle will lead to the side that has the most negative camber.
   D. If the camber angle is out of specification, this may be caused by worn suspension components.

8. Technician A says some abnormal tire wear conditions are normal for certain design tires. Technician B says the TMC Radial Tire Conditions Analysis Guide or Manufacturers Tread Wear Reference Guide should be used to determine tire wear pattern conditions. Who is correct?
   A. A only
   B. B only
   C. Both A and B
   D. Neither A nor B

2013 © Delmar, Cengage Learning

9. Referring to the figure above, the position of a sliding fifth wheel is being discussed. Technician A says that if the fifth wheel is too far ahead, the increased weight transfer to the steer axle might increase steering effort. Technician B says if the fifth wheel is too far rearward, there is not enough weight transfer to the steering axle causing premature front-wheel lockup during a hard brake application. Who is correct?
   A. A only
   B. B only
   C. Both A and B
   D. Neither A nor B

10. How should steering knuckle vertical clearance be adjusted?

    A. Adding or subtracting shims
    B. Selecting the correct draw key
    C. Using the appropriate bushings
    D. Adjusting a stop bolt and jam nut

11. Of the following scenarios, which one would be the LEAST LIKELY to require a steering sensor recalibration to be performed?

    A. Maintenance or repair work on the steering linkage, steering gear, or other related mechanism
    B. Wheel alignment or wheel track adjustment
    C. Steering wheel replacement
    D. Tire and wheel on vehicle balancing

12. Two technicians are discussing power steering pump damage. Technician A says on disassembly that score marks on the pump drive gear indicate damage. Technician B says that if an overheating condition exists, you must disassemble a power steering pump to determine the extent of any damage. Who is correct?

    A. A only
    B. B only
    C. Both A and B
    D. Neither A nor B

13. All of the following statements about negative caster are true EXCEPT:

    A. Negative caster does not help return the front wheels to the straight-ahead position after a turn.
    B. Negative caster contributes to directional instability and reduced directional control.
    C. Negative caster contributes to front-wheel shimmy.
    D. Negative caster reduces road shock transmitted to the suspension and chassis.

14. A driver complains that after his vehicle is driven on a stretch of rough road, the steering wheel continues to shake for a few seconds. This could be caused by:

    A. Leaking front shock absorbers.
    B. Low tire pressure.
    C. A rusted rear shock absorber.
    D. A missing jounce bumper.

15. Technicians are discussing wheel and tire balance procedures. Technician A says that dynamic wheel balancing is performed with the wheel mounted on the axle. Technician B says that prior to removal from the hub, the position of the dynamically balanced tires should be noted and the tire/wheel assembly should be installed in the same position. Who is correct?

    A. A only
    B. B only
    C. Both A and B
    D. Neither A nor B

16. Which of the following would cause a featheredged wear pattern across the tire tread?
    - A. Incorrect toe setting
    - B. Incorrect camber setting
    - C. Incorrect caster setting
    - D. Incorrect tire pressure

17. What kind of spindle does a conventional design front axle beam hub use?
    - A. Tapered
    - B. Straight
    - C. Integral
    - D. Unitized

18. A driver complains of high turning effort at idle on a vehicle equipped with power steering. Above idle RPM, the system operates normally. What is the LEAST LIKELY cause of this complaint?
    - A. An unobstructed fluid supply/pressure line
    - B. Excessive wear between vanes and rotor of the power steering pump assembly
    - C. A sticking flow control valve
    - D. An obstruction or kink in the fluid return line

19. When discussing shock absorbers, all of the statements are true EXCEPT:
    - A. Shock absorbers control spring action and body sway.
    - B. Shock absorbers help maintain tire tread contact on the road.
    - C. Shock absorbers provide more control on the compression cycle than on the extension cycle.
    - D. Shock absorbers improve vehicle handling and steering control.

20. Steering wheel shimmy at high speeds may be caused by:
    - A. Too much positive caster.
    - B. Out-of-balance wheel and tire assemblies.
    - C. Air in the steering system.
    - D. Low tire pressure.

21. Load distribution is being discussed. Technician A says that if the fifth wheel is too far ahead, more weight is transferred to the steer axle increasing steering effort. Technician B says if the fifth wheel is too far rearward, not enough weight will be transferred to the steer axle causing steering instability. Who is correct?
    - A. A only
    - B. B only
    - C. Both A and B
    - D. Neither A nor B

22. The LEAST LIKELY effect of misaligned rear drive axles is:

    A. Steering wheel returning much too fast after cornering.
    B. Steering pull.
    C. Abnormal tire wear.
    D. Erratic steering.

23. The driver complains of a popping sound and a shake in the steering above 45 mph. What is the most likely cause?

    A. Worn shock absorber bushings
    B. Out-of-balance tire
    C. Incorrect toe-in setting
    D. Separated belt in the tire

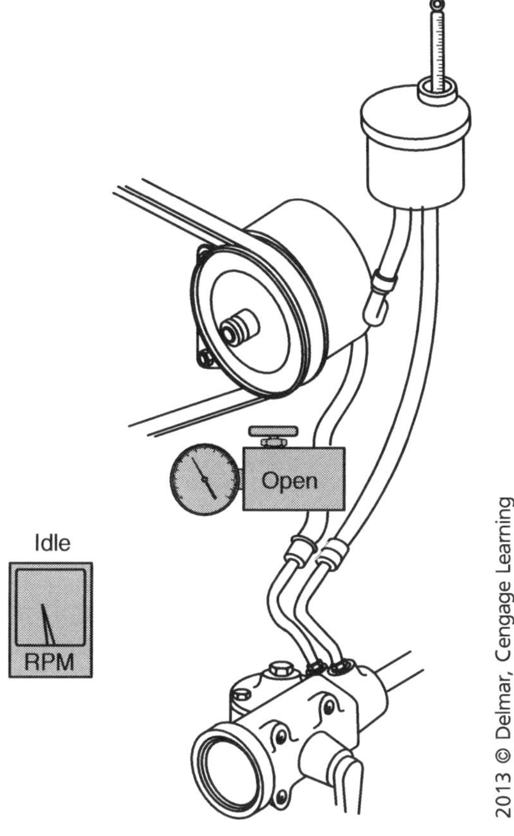

2013 © Delmar, Cengage Learning

24. Referring to the figure above, two technicians are discussing the procedure for diagnosing a steering system using a steering system analyzer. Technician A states that when the engine is at idle and the gate valve on the analyzer is open, if the pressure is less than 200 psi there may be a restriction in the high-pressure hose. Technician B says that when the flow valve in the analyzer is closed and the pressure is increased to the original equipment manufacturer (OEM) specified test value, if the flow in gallons per minute (GPM) is less than the test value specified by the OEM, the pump may be defective. Who is correct?

    A. A only
    B. B only
    C. Both A and B
    D. Neither A nor B

25. Two technicians are discussing shock absorber inspection. Technician A says that if one shock is defective, both shock absorbers across that axle should be replaced. Technician B says that while performing the inspection, if the shock absorbers are not leaking they are still good. Who is correct?

    A. A only
    B. B only
    C. Both A and B
    D. Neither A nor B

26. A technician is inspecting the steering system for excessive steering wheel free-play on a unit with a 22-inch steering wheel. Which of these statements is correct?

    A. Free-play cannot exceed 2 1/4 inches.
    B. Free-play cannot exceed 2 3/4 inches.
    C. Free-play cannot exceed 3 inches.
    D. Free-play cannot exceed 3 1/4 inches.

27. After installing the cross tube assembly, a technician finds that the tie rod clamp interferes with the I-beam on a full turn. If not corrected, which of the following symptoms would result?

    A. A noise when turning over bumps
    B. Damaged drag link assembly
    C. Abnormal steering control
    D. Inaccurate camber measurement

28. What is the LEAST LIKELY benefit that wheel bearing lubrication provides?

    A. Prevents contamination
    B. Cools the bearings
    C. Cleans the bearings
    D. Prevents corrosion

29. Two technicians are discussing trailer air suspension systems with an external dock lock mechanism. All of the statements are true EXCEPT:

    A. During normal operation, air pressure from the trailer air supply system is supplied to an EDL chamber.
    B. When the trailer parking brake is applied, air pressure is released from the EDL chamber.
    C. When the flip plates are in the upward position, they are in contact with the air suspension system.
    D. The flip plates maintain the trailer at dock height and prevent the trailer from walking forward.

30. Worn or bent steering column components may cause all of the following EXCEPT:

    A. A rattling noise.
    B. Excessive steering free-play.
    C. A binding condition when turning the steering wheel.
    D. Improper front suspension alignment.

31. While discussing camber alignment angles and their effect on a heavy-duty tractor, Technician A says that camber should be set slightly positive. Technician B says that setting camber slightly positive will compensate for normal deflection of the axle and front suspension. Who is correct?

    A. A only
    B. B only
    C. Both A and B
    D. Neither A nor B

32. When discussing lift axles with air suspension and air lift, all of the following statements are true EXCEPT:

    A. Two quick-release valves are connected in the air system.
    B. The air system incorporates a pressure protection valve.
    C. Air pressure to the suspension air bags may be adjusted.
    D. When the manual valve or toggle switch is in the on position, air is supplied to the lift air bags to raise the axle.

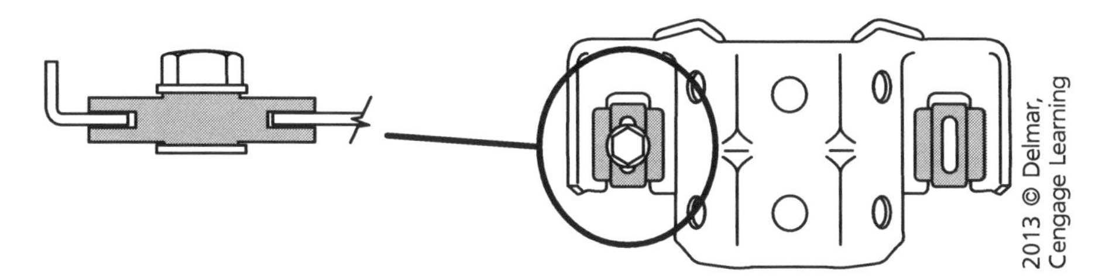

33. Referring to the figure above, two technicians are talking about inspection of collapsible steering columns. Technician A says to check the contact between the bolt head and the bracket. If the bolt head contacts the bracket, the shear load is too high and the column must be replaced. If the measurement between the bolt head and the bracket is not within specifications, replace the column. Technician B says to measure the clearance between the capsules and the slots in the steering column bracket. If this measurement is not within specifications, replace the bracket.

    A. A only
    B. B only
    C. Both A and B
    D. Neither A nor B

34. While performing a front-end alignment, a technician notes that the caster angle on the left side of the vehicle is 2 degrees and on the right side of the vehicle the caster angle is 4 degrees. Technician A says that this might indicate that the front axle is twisted. Technician B says that it is possible that the caster shims underneath the springs are of unequal thickness. Who is correct?

    A. A only
    B. B only
    C. Both A and B
    D. Neither A nor B

35. When the pintle hook is closed, the pintle hook safety latch is held in the lock position:

    A. Electrically.
    B. By air pressure.
    C. By spring tension.
    D. Manually by a lever.

36. The LEAST LIKELY cause of a bowed frame rail is:

    A. Operating a dump truck with the box up and loaded.
    B. Snow plow operation.
    C. Vehicle overload.
    D. Unequal loading of the frame.

37. The LEAST LIKELY affect of positive caster is:

    A. Positive castor reduces steering effort while turning.
    B. Correct positive caster provides improved directional stability of the vehicle.
    C. Excessive positive caster produces harsh riding quality.
    D. When there is a difference between the caster setting for the left and right front tires, the vehicle will pull to the most positive side.

38. What connects a pitman arm to the steering control arm?

    A. Drag link
    B. Tie rod assembly
    C. Steering knuckle
    D. Ackerman arm

39. When adjusting power steering gears, Technician A states that the wormshaft bearing preload adjustment should be checked with a torque wrench while rotating the gear stub shaft. Technician B states that the pitman shaft over center preload adjustment should be performed with the intermediate steering shaft connected to the steering gear. Who is correct?

    A. A only
    B. B only
    C. Both A and B
    D. Neither A nor B

40. Technician A says a bent pitman arm will cause toe-out on turns to be incorrect. Technician B says a bent tie rod will cause an incorrect turning angle. Who is correct?

    A. A only
    B. B only
    C. Both A and B
    D. Neither A nor B

41. Two technicians are discussing bearing characteristics. All of the statements concerning bearings are true EXCEPT:

    A. Cylindrical ball bearings are designed primarily to withstand radial loads, but these bearings may also handle considerable thrust loads.

    B. Cylindrical roller bearings are designed primarily to carry radial loads, but they can also handle some thrust load.

    C. Tapered roller bearings have excellent radial, thrust, and angular load-carrying capabilities.

    D. Needle roller bearings are very compact, and they are designed to carry thrust loads. They will not carry radial loads.

42. A driver complains that the air slide release fifth wheel is hard to release. Technician A says that raising the landing gear to relieve pressure on the plungers should help. Technician B says that if the locking plungers will not release, the problem may be a defective air cylinder. Who is correct?

    A. A only

    B. B only

    C. Both A and B

    D. Neither A nor B

43. While performing a front-end alignment, a technician finds that the left-front wheel on an I-beam front suspension has excessive negative camber. This may be caused by:

    A. Improper shim thickness between the underside of the spring and the axle.

    B. Worn knuckle pin bushings on the left end of the axle.

    C. Seized bearing between the steering knuckle and the axle eye.

    D. Worn rear spring shackles and pins.

44. Tire matching is being discussed. Technician A says excessive tire wear may be a result of mismatching. Technician B says mismatched tire sizes across a drive axle may cause high axle lubricant temperature. Who is correct?

    A. A only

    B. B only

    C. Both A and B

    D. Neither A nor B

45. Erratic torque once every 360 degrees through mesh while rotating the steering wheel is caused by:

    A. A bent worm gear in the steering gear box.

    B. Loose wheel lug nuts.

    C. Premature tie rod wear.

    D. Constant drag when turning the steering wheel.

46. Two technicians are discussing front-wheel caster angles. Technician A says a vehicle has better directional control when the front axle caster angle is set positive. Technician B says that when the front axle positive caster angle is increased, the ride quality of the vehicle improves. Who is correct?

    A. A only
    B. B only
    C. Both A and B
    D. Neither A nor B

47. A vehicle is found to be "dog tracking." Upon inspection, the technician finds the center bolt broken in the spring assembly. Technician A says that worn or loose spring shackles could be the cause. Technician B says that loose U-bolts could be the cause. Who is correct?

    A. A only
    B. B only
    C. Both A and B
    D. Neither A nor B

48. Gold/yellow color wheel studs are used in what hub assembly?

    A. All hub-piloted 10-stud hub assemblies.
    B. All hub-piloted eight-stud hub assemblies
    C. All right-hand ball-seat nut hub assemblies
    D. All left-hand ball-seat nut hub assemblies.

49. Which statement is correct concerning turning radius (toe-out on turns)?

    A. Turning radius is adjusted by lengthening or shortening the tie rod.
    B. The stop bolts are adjusted to limit it.
    C. Turning radius will affect tire tread wear.
    D. The inside wheel is parallel to the outside wheel when cornering.

50. When performing an axle alignment on a vehicle equipped with an air ride suspension, Technician A states that the frame height should be checked prior to performing the alignment. Technician B says that the rear axle should be aligned to the steer axle. Who is correct?

    A. A only
    B. B only
    C. Both A and B
    D. Neither A nor B

SECTION

6 Answer Keys and Explanations

INTRODUCTION

Included in this section are the answer keys for each preparation exam, followed by individual, detailed answer explanations and a reference identifying the designated task area being assessed by each specific question. This additional reference information may prove useful if you need to refer back to the task list located in Section 4 of this book for additional support.

PREPARATION EXAM 1 – ANSWER KEY

1. D
2. B
3. C
4. B
5. D
6. A
7. C
8. B
9. A
10. A
11. D
12. A
13. D
14. A
15. B
16. B
17. A
18. C
19. A
20. A
21. A
22. C
23. C
24. D
25. D
26. A
27. D
28. C
29. A
30. C
31. C
32. C
33. A
34. B
35. C
36. B
37. C
38. B
39. D
40. D
41. C
42. D
43. B
44. C
45. D
46. B
47. C
48. D
49. A
50. B

PREPARATION EXAM 1 – EXPLANATIONS

TASK B.3

1. A device that is used on suspensions to primarily absorb energy and dampen suspension oscillation is called:

   A. An equalizer bracket.

   B. A torque rod.

   C. A spring.

   D. A shock absorber.

Answer A is incorrect. An equalizer bracket is used to mount and house the equalizer.

Answer B is incorrect. The torque arm is used to control axle rotational forces and to maintain and adjust axle alignment.

Answer C is incorrect. The spring assemblies distribute the vehicle load across the frame and provide a cushion between the road surface and the vehicle.

Answer D is correct. Shock absorbers dampen spring oscillation by limiting the effects of spring jounce and rebound.

TASK A.5

2. Upon inspection, excessive play in the steering column assembly on a truck equipped with a tilt steering column is noted. Which of the following would be the LEAST LIKELY cause?

   A. Loose lock shoe in the support

   B. Faulty anti-lash spring in the centering sphere

   C. Loose tilt head pivot pins

   D. Column mounting bracket bolts loose

Answer A is incorrect. Tilt mechanisms are designed to allow for the column to be adjusted and then held securely in that position by the lock shoe mechanism. If the lock shoe pin and wedge lock components are worn, excessive movement in the steering column would be the result.

Answer B is correct. A faulty anti-lash spring might cause the housing to be loose, but it would not cause excessive play in the steering column assembly.

Answer C is incorrect. The main housing pivots on the pivot screws located on each side of the main housing support bracket. If the pins become worn, the steering column would have excessive movement because the pivot pins are not holding the column securely in place.

Answer D is incorrect. The column mounting bolt bracket secures the column. If the bracket bolts become loose, the column would shift and have movement in the steering column assembly.

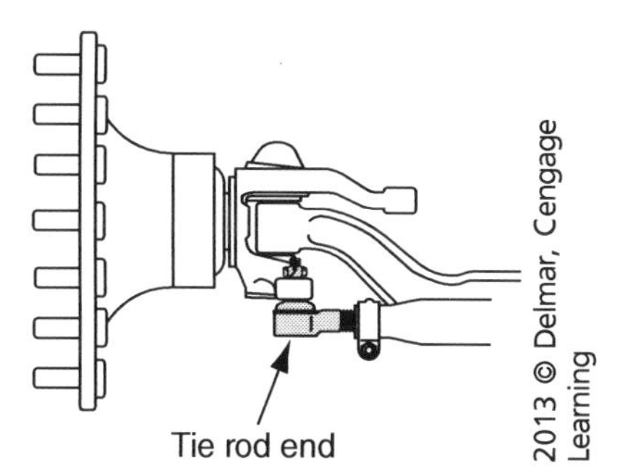

3. Referring to the figure above, when checking a tie rod end for wear which procedure is correct?

TASK A.19

   A. Use a 2 × 4 to pry on the cross tube.
   B. Use a dial indicator and hydraulic jack.
   C. Use up and down hand force to check for wear.
   D. Use a feeler gauge to check for wear.

   Answer A is incorrect. Excessive leverage might be applied using a 2 × 4 and might damage the tie rod socket.

   Answer B is incorrect. A dial gauge is used to measure wear, but a hydraulic jack should never be used to check for movement.

   Answer C is correct. Hand pressure should only be used to check for movement.

   Answer D is incorrect. Wear is measured with a dial indicator, not a feeler gauge.

4. While discussing steering system components, Technician A states that a typical single-axle truck system has one Ackerman arm. Technician B says that the Ackerman arm controls turning radius. Who is correct?

TASK C.6

   A. A only
   B. B only
   C. Both A and B
   D. Neither A nor B

   Answer A is incorrect. There are two Ackerman arms in the steering system.

   Answer B is correct. Only Technician B is correct. Ackerman arms control the turning radius and allow for each front wheel to rotate on different arcs while turning.

   Answer C is incorrect. Only Technician B is correct.

   Answer D is incorrect. Technician B is correct.

5. A driver complains that after hitting a bump his vehicle suddenly veers to the right or left. Which of these is the LEAST LIKELY cause?

TASK C.1

   A. A loose idler arm
   B. A damaged relay rod
   C. A worn tie rod end
   D. A wheel out of balance

   Answer A is incorrect. A loose idler arm will cause fluctuation in the steering when striking a bump.

   Answer B is incorrect. A damaged relay rod can cause veering when striking a bump because of looseness in the ball joint end of the relay rod.

   Answer C is incorrect. A worn tie rod end can cause veering when striking a bump because of looseness in the ball joint end of the tie rod.

   Answer D is correct. Wheel balance will not cause veering, only a shaking in the steering wheel.

TASK D.1

6. The alignment angle that is LEAST LIKELY to cause the greatest tire wear is:

   A. Castor.
   B. Turning radius.
   C. Toe-in.
   D. Camber.

Answer A is correct. Castor is a non-tire-wearing angle. Excessive castor will be present with steerability complaints and wheel shimmy at low speeds.

Answer B is incorrect. Turning radius usually does not cause tire wear unless the Ackerman arms are damaged or the wrong Ackerman arm is installed.

Answer C is incorrect. Toe-in causes the greatest tire wear. Improper toe adjustment causes rapid tire wear, which may result in tire failure, collision damage, and personal injury. Excessive toe-out causes wear on the inside of the tire tread ribs and a sharp feathered edge on the outside of the tread ribs. If excessive toe-in is present, the tire tread wear is reversed.

Answer D is incorrect. Camber can cause tire wear but not as great a tire wearing angle as toe-in.

TASK A.2

7. Technician A states that some steering shaft universal joints have different splines on the upper and lower yokes. Technician B says that when installing the universal joint, install it on the lower end of the intermediate shaft first. Who is correct?

   A. A only
   B. B only
   C. Both A and B
   D. Neither A nor B

Answer A is incorrect. Technician B is also correct.

Answer B is incorrect. Technician A is also correct.

Answer C is correct. Both Technicians are correct. Some steering shaft universal joints have different splines on the upper and lower yokes. Be sure the yoke splines and shaft splines are matched. Also, when installing the universal joint, install it on the lower end of the intermediate shaft first. Then install the U-joint on the steering gear stub shaft.

Answer D is incorrect. Both Technicians are correct.

TASK B.8

8. Technician A states that in order for a steerable suspension to steer or track correctly, it is necessary for the front wheels to be in a toe-out condition. Technician B says that the toe setting should be between 1/16 inch and 1/8 inch. Who is correct?

   A. A only
   B. B only
   C. Both A and B
   D. Neither A nor B

Answer A is incorrect. The front wheels need to be in a toe-in condition.

Answer B is correct. Only Technician B is correct. The toe setting should be between 1/16 inch and 1/8 inch.

Answer C is incorrect. Only Technician B is correct.

Answer D is incorrect. Technician B is correct.

9. Alignment procedures are being discussed. Technician A says that installing the caster shim between the axle and spring with the thick part to the rear will make the caster angle positive. Technician B says installing the caster shim between the axle and spring with the low side at the front will make the caster angle more negative. Who is correct?

A. A only

B. B only

C. Both A and B

D. Neither A nor B

Answer A is correct. Only Technician A is correct. You make caster more positive by installing a shim between the axle and spring with the high side at the rear.

Answer B is incorrect. You make the caster more negative by installing a shim between the axle and spring with the low side at the rear.

Answer C is incorrect. Only Technician A is correct.

Answer D is incorrect. Technician A is correct.

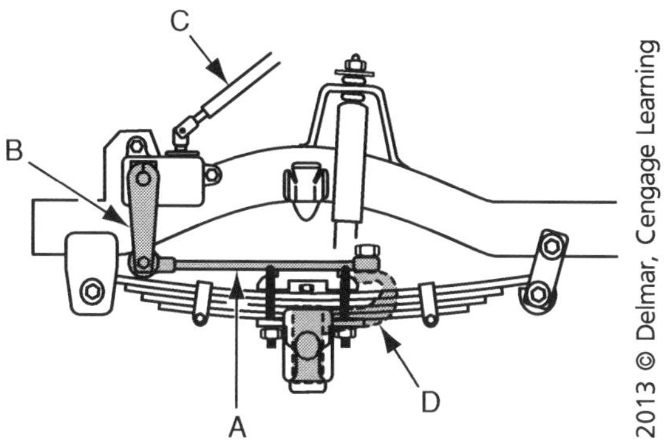

10. Referring to the figure above, the drag link is:

A. Letter A.

B. Letter B.

C. Letter C.

D. Letter D.

Answer A is correct. Letter A is the drag link.

Answer B is incorrect. Letter B is the pitman arm.

Answer C is incorrect. Letter C is the steering shaft.

Answer D is incorrect. Letter D is the steering arm.

11. A driver complains about rough ride. Which of these would be the LEAST LIKELY cause?

A. Excessive positive caster

B. Leaking or damaged shock absorber

C. Loose/worn suspension component

D. Defective rear wheel bearings

Answer A is incorrect. Excessive positive caster can cause rough ride characteristics.

Answer B is incorrect. A leaking or damaged shock absorber can cause rough ride characteristics.

Answer C is incorrect. A worn or loose suspension component can cause rough ride characteristics.

Answer D is correct. Defective rear axle wheel bearings are the exception. Although defective wheel bearings can be dangerous and generate noise and possible wheel wobble, they do not usually create a rough ride.

TASK C.4

12. When a technician is going to check out the front-end geometry of a heavy-duty truck, which of the following steps should be performed first?

    A. Neutralize the suspension by driving the unit back and forth in a straight line.
    B. Jack the front end up and scribe the front tires.
    C. Disconnect the drag link.
    D. Check the over-center adjustment of the steering gear.

    Answer A is correct. The suspension should be neutralized to take any load or torque off of the suspension and steering components.

    Answer B is incorrect. Scribing the tires is part of the toe setting, which is the last adjustment made.

    Answer C is incorrect. It is not necessary to remove the drag link to perform alignment checks.

    Answer D is incorrect. Steering gear adjustments have no bearing on steering geometry.

TASK A.1

13. A driver complains of excessive play in the steering column assembly on a truck equipped with a tilt steering column. All of the following might be the cause EXCEPT:

    A. Loose column mounting bracket bolts.
    B. Loose tilt head pivot pins.
    C. A loose lock shoe pin in the support.
    D. A faulty anti-lash spring in the centering sphere.

    Answer A is incorrect. Loose column mounting bracket bolts might cause excessive play in the steering column assembly. The column mounting bolts should be inspected and tightened to the proper torque.

    Answer B is incorrect. Pivot screws on each side of the main housing allow the main housing to pivot on the support bracket. Worn or loose pins allow for excessive play in the column.

    Answer C is incorrect. The lock shoe pin and tilt mechanisms are designed to allow for the column to be adjusted and then held securely in that position. Worn lock shoe pin and wedge lock components would allow for excessive movement in the steering column.

    Answer D is correct. A faulty anti-lash spring might cause the housing to be loose, but it would not cause excessive play in the steering column assembly.

TASK D.5

14. A truck tire is found to have been run in an underinflated condition. Technician A says the tire may be damaged internally and should be removed from the truck and inspected. Technician B says that a tire should be balanced prior to checking wheel runout. Who is correct?

    A. A only
    B. B only
    C. Both A and B
    D. Neither A nor B

    Answer A is correct. Only Technician A is correct. When a truck tire has been run in a very underinflated condition, you must remove it from the truck, inspect the casing for damage, and properly remount it.

    Answer B is incorrect. Before balancing a tire, the technician should check wheel runout.

    Answer C is incorrect. Only Technician B is correct.

    Answer D is incorrect. Technician B is correct.

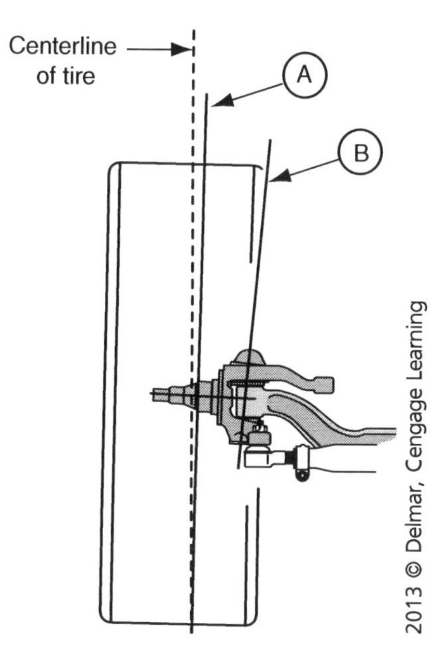

15. Referring to the figure above, what angle is being represented at Letter B?

A. Vertical angle
B. Kingpin inclination
C. Camber
D. Caster

Answer A is incorrect. There is no such angle when talking about front-end alignment angles.

Answer B is correct. Kingpin inclination is the angle formed between a true vertical line and a line drawn through the kingpin axis.

Answer C is incorrect. Camber is the angle formed between a true vertical line and a line drawn through the centerline of the tire.

Answer D is incorrect. Castor is the forward or rearward tilt of the kingpin at the top and is not displayed in the figure.

16. If the power steering gear poppet valves are misadjusted, which of the following conditions may be exhibited?

A. Steering wheel kick
B. Reduced wheelcut
C. Directional pull
D. Non-recovery

Answer A is incorrect. The misadjusted power steering gear poppet valves will not cause steering wheel kick. Wheel kick or bump steer occurs when the steering wheel reacts to a bump that the wheels hit.

Answer B is correct. Misadjusted power steering gear poppet valves will cause reduced wheelcut. Wheelcut occurs when the steering wheel cannot be rotated far enough in a right or left turn. The poppet valves provide a hydraulic pressure relief on the power side of a turn, thus reducing the power assist and misadjusted puppet valves could prevent the wheel from rotating far enough.

Answer C is incorrect. The misadjusted power steering gear poppet valves will not cause directional pull.

Answer D is incorrect. The misadjusted power steering gear poppet valves will not cause non-recovery of the steering.

TASK B.10

17. A vehicle is found to be "dog tracking." Upon inspection, the technician finds no issues with worn components and says the axle alignment should be checked. Which is the LEAST LIKELY method to use for checking axle alignment?

A. Light and laser beam systems
B. Computer-controlled sensor systems
C. Straightedge and tram method
D. String and stick method

Answer A is incorrect. Light and laser beam systems are an acceptable method.

Answer B is incorrect. Computer-controlled sensor systems are an acceptable method.

Answer C is incorrect. Straightedge and tram method is an acceptable method.

Answer A is correct. String and stick method is very inaccurate.

TASK D.1

18. Technician A says excessive load/ weight on a steer axle could not cause abnormal tire wear or steering issues. Technician B says excessive load/ weight on one wheel end of the steer axis could cause abnormal tire wear or steering issues. Who is correct?

A. A only
B. B only
C. Both A and B
D. Neither A nor B

Answer A is incorrect. Technician B is also correct.

Answer B is incorrect. Technician A is also correct.

Answer C is correct. Both Technicians are correct. Steer axle alignment geometry is designed with the axle weight rating in mind. Excessive steer axle weight might change the degree of camber at the front wheel assemblies. This would cause tire wear. Excessive load/weight might also cause steerability issue with increased steering effort. Excessive weight on just one wheel end might cause the vehicle to pull and would also affect the camber angle at that position.

Answer D is incorrect. Both Technicians are correct.

TASK A.4

19. Which of the following is the acceptable method for removing a steering wheel from the steering shaft?

A. Use a steering wheel puller.
B. Firmly grasp with both hands and pull.
C. Use two pry bars and remove.
D. Strike the center shaft with a hammer to drive it out of the steering wheel.

Answer A is correct. The only acceptable method to remove a steering wheel is to use a puller.

Answer B is incorrect. The steering wheel can't be removed using this method.

Answer C is incorrect. Using pry bars may damage other steering column components.

Answer D is incorrect. Striking the shaft with a hammer will damage the shaft and or retaining nut threads.

20. When performing kingpin bushing wear dial indicator checks, what is the wear limit?

TASK B.2

A. 0.010 inches
B. 0.100 inches
C. 0.030 inches
D. 0.025 inches

Answer A is correct. The wear limit is 0.010 inches.

Answer B is incorrect. A measurement of 0.100 is much too high, and would cause noise in the spindle assembly and steerability issues.

Answer C is incorrect. Kingpin and kingpin bushing replacement might be warranted if the indicator reading was 0.030 inches.

Answer D is incorrect. 0.025 inches is over the maximum wear limit specification. Further inspection would be warranted to determine the cause.

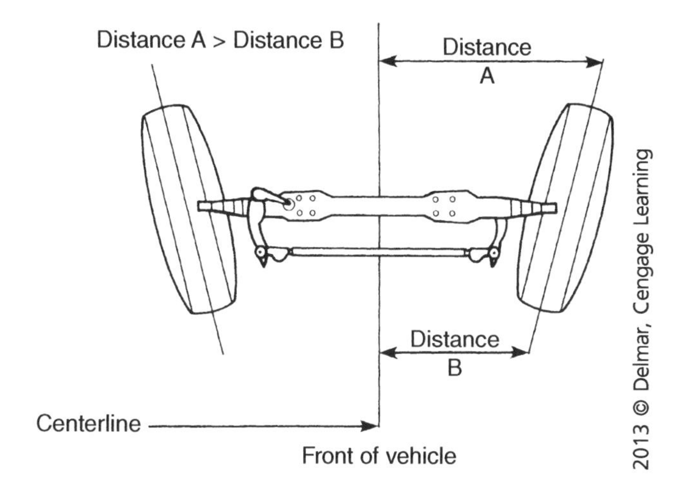

21. Referring to the figure above, what condition is being represented?

TASK C.4

A. Toe-in condition
B. Camber condition
C. Toe-out condition
D. Steering axis inclination

Answer A is correct. The measurement shown in the figure is a toe-in condition.

Answer B is incorrect. Camber is the tilting in or out of the top of the tire as viewed from the front.

Answer C is incorrect. Toe-out would have the rear of the tires closer together than the front.

Answer D is incorrect. Steering axis inclination is the angle formed between the steering axis and a true vertical.

TASK C.5

22. Technician A says when aligning the rear axle position on some spring designs the upper and lower torque rod collars are rotated. Technician B says you align some rear axles by rotating eccentric bushings in the spring hangers. Who is correct?

   A. A only
   B. B only
   C. Both A and B
   D. Neither A nor B

Answer A is incorrect. Technician B is also correct.

Answer B is incorrect. Technician A is also correct.

Answer C is correct. Both Technicians are correct. Some rear suspension torque rods require rotating collars to lengthen or shorten the rod for correct axle placement while others require the rotation of eccentric bushings to locate the axle.

Answer D is incorrect. Both Technicians are correct.

TASK D.9

23. A Technician has found a tire that is under 80 percent of the recommended inflation pressure. Technician A says that before removing the tire and wheel assembly from the vehicle, the technician should deflate the tire by removing the valve core completely. Technician B says that once the tire is deflated and removed it should be placed in a restraining device to check for leaks. Who is correct?

   A. A only
   B. B only
   C. Both A and B
   D. Neither A nor B

Answer A is incorrect. Technician B is also correct.

Answer B is incorrect. Technician A is also correct.

Answer C is correct. Both Technicians are correct. This is an OSHA requirement. A tire that is found to beat or under 80 percent of the recommended tire inflation pressure is considered flat by OSHA standards. Removing the valve stem and deflating the tire prior to removal ensures that there is no possibility of tire/wheel separation. Because it is considered to be flat, the tire should be placed in a tire restraining device and aired using a clip-on air chuck and an extended air hose equipped with an inline gauge to prevent the Technician from being injured in the event of a tire/wheel separation event.

Answer D is incorrect. Both Technicians are correct.

24. A vehicle leans to one side. Which of these is the LEAST LIKELY cause?

TASK B.12

A. One or more broken leaf springs
B. A weak or fatigued spring assembly
C. Unmatched spring design/load capacity spring assemblies
D. A bent or twisted frame rail

Answer A is incorrect. Broken leaf springs are a common cause of leaning to one side.

Answer B is incorrect. Weak or fatigued spring assemblies will not carry the vehicle load. The vehicle leaning to one side is a common occurrence if only one leaf spring assembly is replaced across an axle.

Answer C is incorrect. If the spring assembly being installed does not have the same design and load capacity characteristics as the one being replaced, a lean condition might occur, especially under a loaded condition.

Answer D is correct. Although a bent or twisted frame will cause this condition, it is the LEAST LIKELY cause. A bent frame is usually a result of an accident.

25. Two technicians are discussing the procedure for diagnosing a steering system using a steering system analyzer. Which of the following should be true to obtain accurate results?

TASK A.8

A. The engine oil should be at operating temperature.
B. The steering system hydraulic oil should be cold.
C. New steering system hydraulic oil should be used.
D. The steering system hydraulic oil should be above 140°F (60°C).

Answer A is incorrect. The engine oil does not pass through the analyzer.

Answer B is incorrect. The steering system hydraulic oil should be above 140°F before testing.

Answer C is incorrect. Unless the hydraulic oil is contaminated or shows signs of overheating, there is no need to replace it.

Answer D is correct. The operating temperature of the hydraulic system should be above 140°F (60°C).

26. While discussing air suspension height control valves, Technician A says that an improperly adjusted air suspension height control valve can cause an offset in vehicle attitude. Technician B says that before adjusting the height control valve, the air system pressure needs to be above 80 psi. Who is correct?

TASK B.7

A. A only
B. B only
C. Both A and B
D. Neither A nor B

Answer A is correct. Only Technician A is correct. An improperly adjusted air suspension height control valve can cause an offset in vehicle attitude.

Answer B is incorrect. The air system pressure needs to be above 100 psi before performing an adjustment to the height control valve. This will ensure proper air bag inflation and ride height measurement.

Answer C is incorrect. Only Technician A is correct.

Answer D is incorrect. Technician A is correct.

TASK D.10

27. Technician A says packing a wheel bearing with grease before filling the hub assembly with oil is acceptable because it provides the wheel bearing with lubricant while the oil settles and has a chance to run into the bearing. Technician B says this is acceptable because the oil and grease will mix once the hub reaches operating temperature. Who is correct?

    A. A only
    B. B only
    C. Both A and B
    D. Neither A nor B

Answer A is incorrect. Technician B is also incorrect.

Answer B is incorrect. Technician A is also incorrect.

Answer C is incorrect. Both Technicians are incorrect.

Answer D is correct. Both Technicians are incorrect. Packing the bearing with grease inhibits the flow of oil to the bearing. Wheel bearing grease and hub oil may not be compatible with each other.

TASK C.6

28. A bent steering arm was replaced and now the vehicle produces excessive tire side scrub during sharp turns. What could cause this condition?

    A. An incorrectly adjusted camber setting
    B. A lack of kingpin lubrication
    C. An incorrect steering arm
    D. An incorrect toe setting

Answer A is incorrect. Incorrect camber would produce tire tread edge wear and possibly cause a pull, but it would not create side scrub.

Answer B is incorrect. A lack of kingpin lubricant could produce steering difficulties and stiffness when turning but not tire side scrub.

Answer C is correct. An incorrect steering arm could affect the Ackerman angle or toe-out on turns, which would change the tire paths during a turn producing tire side scrub.

Answer D is incorrect. An incorrect toe setting would produce scrubbing and featheredge wear at all times.

TASK B.2

29. A driver complains of a front end shimmy with slight vibrations. Technician A says loose or worn kingpins or kingpin bushings might be the cause. Technician B says that an overloaded trailer might be the cause. Who is correct?

    A. A only
    B. B only
    C. Both A and B
    D. Neither A nor B

Answer A is correct. Only Technician A is correct. A front shimmy with slight vibrations could be caused by a loose kingpin or kingpin bearing. Some OEMs refer to the kingpin as a knuckle pin.

Answer B is incorrect. Overloading will not cause front end shimmy.

Answer C is incorrect. Only Technician A is correct.

Answer D is incorrect. Technician A is correct.

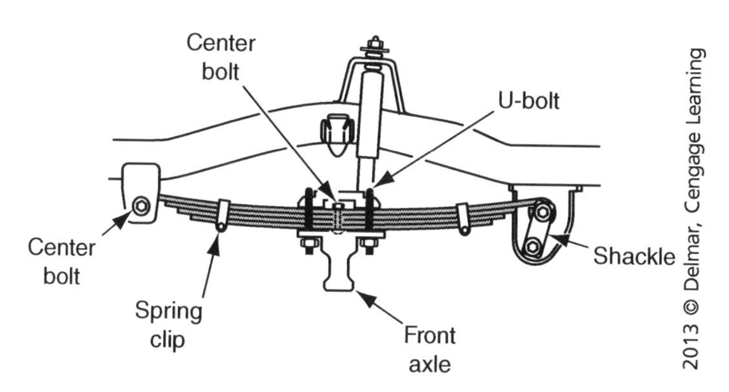

30. Referring to the figure above, what component provides self-dampening properties to the leaves in a spring pack?

TASK B.4

A. Spring clips
B. U-bolts
C. Center bolt
D. Spring pins

Answer A is incorrect. Spring clips limit the amount of separation that can occur between the leaves.

Answer B is incorrect. U-bolts are used as the clamping/mounting component for the spring assembly and the axle.

Answer C is correct. The tension that the center bolt loads the spring leaves under helps define the self-dampening ability of the spring assembly.

Answer D is incorrect. The spring pins are used to retain the leaf spring assembly in the spring hangers.

31. A technician is inspecting a hub assembly and finds two broken studs. The technician should:

TASK D.3

A. Remove and replace the broken studs plus the adjacent studs.
B. Remove the broken studs plus half of the remaining studs.

C. Remove and replace all of the wheel studs.
D. Only replace the two broken studs.

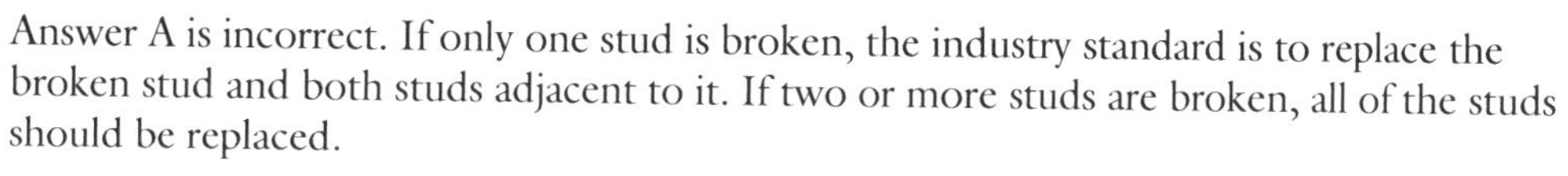
Answer A is incorrect. If only one stud is broken, the industry standard is to replace the broken stud and both studs adjacent to it. If two or more studs are broken, all of the studs should be replaced.

Answer B is incorrect. Multiple broken studs on a hub is an indication that an issue is present at the hub. This condition may have been caused by improper mounting, inadequate torque, or overload. All of the studs should be replaced due to the stresses that were placed on all of the studs.

Answer C is correct. All of the studs should be replaced due to the stresses that were placed on all of the studs. If only the broken studs and those adjacent to them are replaced, the remaining studs may break due to previous stress factors.

Answer D is incorrect. Multiple broken studs on a hub is an indication that an issue is present at the hub. This condition may have been caused by improper mounting, inadequate torque, or overload. All of the studs should be replaced due to the stresses that were placed on all of the studs.

TASK C.1

32. All of the following can cause abnormal tire wear, shimmy, or vibration EXCEPT:

    A. Tire/wheel imbalance.
    B. Excessive wheel or hub runout.
    C. Excessive cam brake stroke.
    D. Improper tire mounting.

Answer A is incorrect. Tire/wheel imbalance may cause a shimmy or vibration.

Answer B is incorrect. Excessive wheel or hub runout may cause a shimmy or vibration.

Answer C is correct. Excessive cam brake stroke can cause poor brake application and air brake imbalance, but it will not cause abnormal tire wear, shimmy, or vibration. Poor brake application and air brake imbalance may result.

Answer D is incorrect. Improper tire mounting may cause a shimmy or vibration.

TASK B.4

33. Upon inspection, a technician finds a single leaf broken within the spring pack. Which of the following is the LEAST LIKELY cause?

    A. The technician would replace only the broken leaf.
    B. The technician would make a careful inspection of the other leaves in that spring pack.
    C. The technician would replace the broken leaf and the same leaf in the other spring assembly on that axle.
    D. The technician would replace both the spring pack with the broken leaf and the other spring pack on that axle.

Answer A is correct. It is acceptable to replace just the broken leaf in a spring pack assembly. However, this would be done only after a careful inspection of the other leaves in the spring pack is done, and generally, best practice is to replace the same leaf in the other spring assembly on the axle.

Answer B is incorrect. Careful inspection of the other leaves should be made to ensure the integrity of the steel is maintained prior to making a decision to replace the broken leaf only.

Answer C is incorrect. Anytime a single leaf is replaced within a spring pack, it is suggested that the same leaf be replaced in the other spring assembly on that axle. This ensures that the same dampening and load characteristics are maintained.

Answer D is incorrect. If upon inspection of the spring pack leaves it is determined that the remaining leaves are not in good condition, the spring pack should be replaced. As a rule, most OEMs recommend replacement of the spring pack anytime a leaf is broken. If the spring pack is replaced, it is a good practice to replace the other spring pack on that axle.

34. A driver complains that he notices a vibration in the steering wheel while traveling at 28 mph. Of the conditions listed below, which is the LEAST LIKELY cause?

TASK D.2

A. Improper wheel balance

B. Worn kingpin bushings

C. Bent wheel mounting surface

D. Shifted belt inside tire

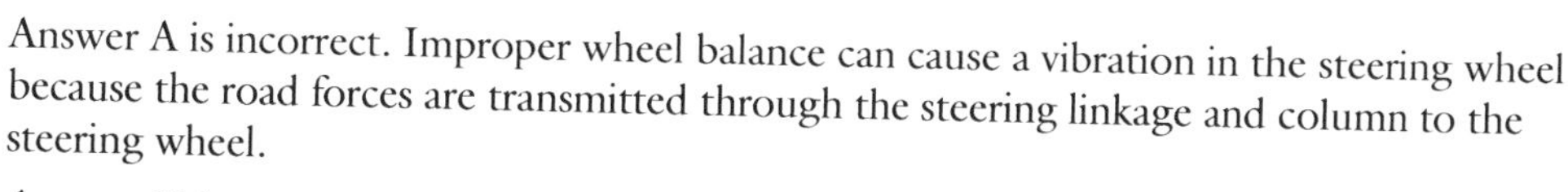

Answer A is incorrect. Improper wheel balance can cause a vibration in the steering wheel because the road forces are transmitted through the steering linkage and column to the steering wheel.

Answer B is correct. Worn kingpin bushings can cause wander and weave issues or darting effects, but are not a likely cause of vibration in the steering wheel.

Answer C is incorrect. A bent wheel mounting surface can cause a dynamic imbalance leading to improper wheel balance and the resulting wheel vibration.

Answer D is incorrect. A shifted belt inside a tire can cause a vibration in the steering wheel. A separated belt forms a gap under the tread causing a popping sound and the separating causes the wheel to shake.

35. All of the following are a function of the torque arm EXCEPT:

TASK B.5

A. To retain axle alignment.

B. To control axle torque.

C. To control ride height.

D. To adjust axle alignment.

Answer A is incorrect. Retaining axle alignment is a function of the torque arm.

Answer B is incorrect. Controlling axle torque is a function of the torque arm.

Answer C is correct. Controlling ride height is not a function of the torque arm.

Answer D is incorrect. Adjusting axle alignment is a function of the torque arm.

36. While discussing wheel balance, Technician A says that dynamic wheel balance is better than static balance because it is performed with the wheel stationary. Technician B says that dynamic wheel balance is better than static balance because you can tell if the wheel is bent. Who is correct?

TASK D.7

A. A only

B. B only

C. Both A and B

D. Neither A nor B

Answer A is incorrect. Dynamic balancing is performed with the tire rotating.

Answer B is correct. Only Technician B is correct. Dynamic balancing can detect a bent wheel rim. When a dynamically unbalanced wheel is rotating, the heavy spot moves around the tire causing lateral wheel shake or shimmy.

Answer C is incorrect. Only Technician B is correct.

Answer D is incorrect. Technician B is correct.

TASK B.6

37. Technician A says that when reinstalling rubber-bushed equalizing beams, you should install the torque rods before torquing the beam end bolts to specification. Technician B says that you should ensure the beams are parallel to the frame rails. Who is correct?

A. A only

B. B only

C. Both A and B

D. Neither A nor B

Answer A is incorrect. Technician B is also correct.

Answer B is incorrect. Technician A is also correct.

Answer C is correct. Both Technicians are correct. Installing the torque arms prior to tightening the beam end bolts allows the beam to move slightly if needed and does not load up the suspension. The beams should be parallel to the frame frail so the suspension tracks correctly.

Answer D is incorrect. Both Technicians are correct.

TASK C.1

38. Which of the following is the most likely cause of steering wheel shimmy?

A. A toe-in setting that is too high

B. Improper dynamic wheel balance

C. An excessive load in the vehicle

D. A toe-out setting that is too high

Answer A is incorrect. A toe-in setting that is too high will cause excessive tire wear.

Answer B is correct. Improper dynamic wheel balance can cause steering wheel shimmy. When the wheel rotates until the heavy spot is at the front of the wheel, the heavy spot movement turns the left-front wheel outward, causing lateral wheel shake or shimmy.

Answer C is incorrect. An excessive vehicle load might cause a wander and weave condition because not enough weight is transferred to the steer axle, but will not cause wheel shimmy.

Answer D is incorrect. A toe-out setting that is too high cannot cause steering wheel shimmy. Toe only affects directional stability and tire wear.

TASK A.10

39. A technician has checked the pulley alignment on a power steering system and found them to be misaligned. All of the following could cause the power steering pump pulley to become misaligned EXCEPT:

A. An overpressed pulley.

B. Loose fit from pulley hub to pump shaft.

C. A worn or loose pump mounting bracket.

D. A broken engine mount.

Answer A is incorrect. An overpressed pulley could cause the power steering pump pulley to become misaligned.

Answer B is incorrect. A loose fit from the pulley hub to the pump shaft could cause the power steering pump pulley to become misaligned.

Answer C is incorrect. A worn or loose pump mounting bracket could cause the power steering pump pulley to become misaligned.

Answer D is correct. A broken engine mount will not cause the power steering pump pulley to become misaligned. All of the components would still move in unison; the whole engine assembly would move.

40. The unitized wheel end may be described as:

TASK D.11

A. Easily identifiable by the two distinct star shapes embossed on the center of its hubcaps.
B. An enclosed unit with bearings that will need to be serviced every 100,000 miles.
C. An enclosed unit with bearings that will need to be serviced every 500,000 miles.
D. An enclosed unit with bearings lubricated for the life of the hub, bearing, and seal assembly.

Answer A is incorrect. The embossed shapes on the hubcaps are half moon shapes.

Answer B is incorrect. Unitized wheel end components are not serviceable. If a failure occurs, the hub assembly will need to be replaced.

Answer C is incorrect. Unitized wheel ends are an enclosed unit with bearings lubricated for the life of the hub, bearing, and seal assembly.

Answer D is correct. Unitized wheel ends are designed for the bearings to run the life of the hub. Periodic inspections should be performed to ensure smooth operation.

41. A truck that is equipped with an air suspension system has one side rising after unloading. Technician A says that this could be caused by a malfunctioning leveling valve. Technician B says that this could be caused by a plugged exhaust port. Who is correct?

TASK B.9

A. A only
B. B only
C. Both A and B
D. Neither A nor B

Answer A is incorrect. Technician B is also correct.

Answer B is incorrect. Technician A is also correct.

Answer C is correct. Both Technicians are correct. An air suspension system that has one side rising after unloading can be caused by a malfunctioning leveling valve or a plugged exhaust port. On a system with dual leveling valves, if one valve is not adjusted properly, this condition may occur. Also, if an exhaust port is plugged, air pressure in the bag will not exhaust and this condition will occur.

Answer D is incorrect. Both Technicians are correct.

42. All of the following can cause hard steering EXCEPT:

TASK C.1

A. A dry fifth wheel.
B. An overloaded steer axle.
C. A contaminated power steering system.
D. Leaking shock absorbers.

Answer A is incorrect. If the fifth wheel is dry the trailer can't pivot properly. This erratic pivoting action would cause hard steering.

Answer B is incorrect. Overloading of the steering axle exerts excess pressures on the steering components and tire/wheel assemblies. The result is hard steering.

Answer C is incorrect. Contaminants in the power steering system may adversely affect the system operation, which would cause hard steering.

Answer D is correct. Leaking shock absorbers would cause drivability issues such as wheel hop but will not cause hard steering.

TASK A.3

43. Two technicians are discussing air cab suspensions. Which of the following statements is correct?

A. It has three springs.

B. It has a leveling valve.

C. It is used with rubber cab mounts.

D. It does not have shock absorbers.

Answer A is incorrect. A cab air suspension does not use springs.

Answer B is correct. It has a leveling valve. The leveling valve maintains the proper air pressure in the air springs to provide the correct cab height. Some cab air suspension systems have a single air spring and shock absorber.

Answer C is incorrect. A cab air suspension does not use rubber cab mounts.

Answer D is incorrect. A cab air suspension uses shock absorbers.

TASK C.1

44. Technician A says a steer axle vibration or shimmy could be caused by steer tire or wheel issues. Technician B says replacing the front tires/wheels from another vehicle with no vibration or shimmy is a viable diagnostic procedure. Who is correct?

A. A only

B. B only

C. Both A and B

D. Neither A nor B

Answer A is incorrect. Technician B is also correct.

Answer B is incorrect. Technician A is also correct.

Answer C is correct. Both Technicians are correct. Steer axle vibration or shimmy could be caused by steer tire or wheel issues, worn steering components, or worn suspension components. Replacing the front tires/wheels from another vehicle with no vibration or shimmy is a viable diagnostic procedure that would eliminate the tire/wheel assemblies as a possible source.

Answer D is incorrect. Both Technicians are correct.

TASK B.1

45. During a front axle assembly inspection, a technician should check for a twisted axle beam if any of the following conditions exist EXCEPT:

A. The difference in caster angle exceeds 0.5 degrees from side to side.

B. The caster shims in place differ by 1 degree or more.

C. A low-speed shimmy exists, and there is no evidence of looseness elsewhere in the steering system.

D. Excessive tire wear exists.

Answer A is incorrect. The maximum difference from side to side for caster is 0.5 degrees. Any reading over 0.5 degrees would indicate that further inspection of the front axle is necessary to determine if the axle is twisted.

Answer B is incorrect. If the caster shims that are installed vary by 1 degree or more, this would indicate that the axle may be twisted. The technician should never try to correct a twisted axle by varying the caster shim thickness.

Answer C is incorrect. Low-speed shimmy with no evidence of looseness in the steering components indicates that the front axle might be twisted. Caster is designed into the front axle to improve steerability. If the axle is twisted, the caster angles are opposed to each other and would produce this symptom.

Answer D is correct. Caster is a non-tire-wearing angle, so a twisted axle beam will not cause excessive tire wear.

46. During a preventive maintenance inspection, the technician finds that one spring pin will not take grease. The first thing the technician should do is:

TASK B.4

   A. Change the zerk fitting.
   B. Remove the weight from the spring assembly.
   C. Replace the spring pin and bushing.
   D. Heat the spring eye.

   Answer A is incorrect. The first thing that should be done is to remove the weight from the spring assembly and try to grease the assembly in a static position. If the technician is still unable to grease the assembly, then the zerk fitting should be removed and the grease passage in the pin inspected.

   Answer B is correct. Any time the suspension system is being greased, the spring assemblies should be placed in a static position. This will allow for grease to flow around the spring pins.

   Answer C is incorrect. This would be the last resort.

   Answer D is incorrect. You should never heat the spring eyes. This might change the temper in the steel allowing to bending or breakage.

47. Technician A says that before checking the fluid level in the power steering reservoir, you should turn the steering wheel slowly and completely from side to side several times to boost the fluid temperature. Technician B says that foaming in the reservoir indicates low fluid level or air in the system. Who is correct?

TASK A.9

   A. A only
   B. B only
   C. Both A and B
   D. Neither A nor B

   Answer A is incorrect. Technician B is also correct.

   Answer B is incorrect. Technician A is also correct.

   Answer C is correct. Both Technicians are correct. The fluid temp should be above 175°F before checking the fluid level. The temperature should be raised by turning the steering wheel completely from side to side. In addition, foaming in the oil is an indication that there is air in the system and the fluid is aerated. This may lead to overheating and components damage if not repaired.

   Answer D is incorrect. Both Technicians are correct.

48. Kingpin inclination (KPI) may be defined as:

TASK C.2

   A. The forward or rearward tilt of the kingpin at the top.
   B. The tracking angle of the tires from a true straight-ahead track.
   C. The inward or outward tilt of the top of the wheel when viewed from the front of the vehicle.
   D. The amount in degrees that the top of the kingpin inclines away from vertical as viewed from the front of the vehicle.

   Answer A is incorrect. Caster is defined as the forward or rearward tilt of the kingpin at the top.

   Answer B is incorrect. The tracking angle of the tires from a true straight-ahead track is toe.

   Answer C is incorrect. Camber is defined as the inward or outward tilt of the top of the wheel when viewed from the front of the vehicle.

   Answer D is correct. KPI is defined as the amount in degrees that the top of the kingpin inclines away from a true vertical line drawn through the center of the tire and wheel assembly as viewed from the front of the vehicle.

TASK B.7

49. Upon inspection, a technician finds one air bag blown out on a tractor with an air ride suspension. The technician inspects the rest of the suspension system and finds no mechanical or fatigue issues. Technician A says that it is acceptable to replace just that air bag. Technician B says that the axle alignment should be checked after replacement. Who is correct?

    A. A only
    B. B only
    C. Both A and B
    D. Neither A nor B

    Answer A is correct. Only Technician A is correct. It is acceptable to replace one air bag if no mechanical or fatigue issues are found that would contribute to the air bag failure.

    Answer B is incorrect. Axle alignment is not necessary because no mechanical issues were found to cause the air bag failure.

    Answer C is incorrect. Only Technician A is correct.

    Answer D is incorrect. Technician A is correct.

TASK A.16

50. A technician found the splines on the pitman arm worn. After replacement, the technician should do all of the following EXCEPT:

    A. Perform a road test when repairs are completed.
    B. Replace both outer tie rod ends.
    C. Check and correct changes in wheel alignment.
    D. Lube the replacement part after installation.

    Answer A is incorrect. The vehicle should be road tested when the job is completed.

    Answer B is correct. You only replace defective tie rod ends.

    Answer C is incorrect. Whenever a steering component is replaced it is recommended to do a wheel alignment.

    Answer D is incorrect. Any replaced part should be lubricated after installation.

PREPARATION EXAM 2 – ANSWER KEY

1. C
2. B
3. A
4. D
5. C
6. A
7. B
8. B
9. D
10. C
11. C
12. C
13. D
14. C
15. D
16. B
17. C
18. B
19. A
20. B
21. C
22. B
23. B
24. C
25. D
26. D
27. D
28. A
29. B
30. C
31. D
32. C
33. C
34. D
35. A
36. C
37. D
38. C
39. D
40. D
41. C
42. C
43. B
44. B
45. A
46. C
47. D
48. D
49. A
50. C

PREPARATION EXAM 2 – EXPLANATIONS

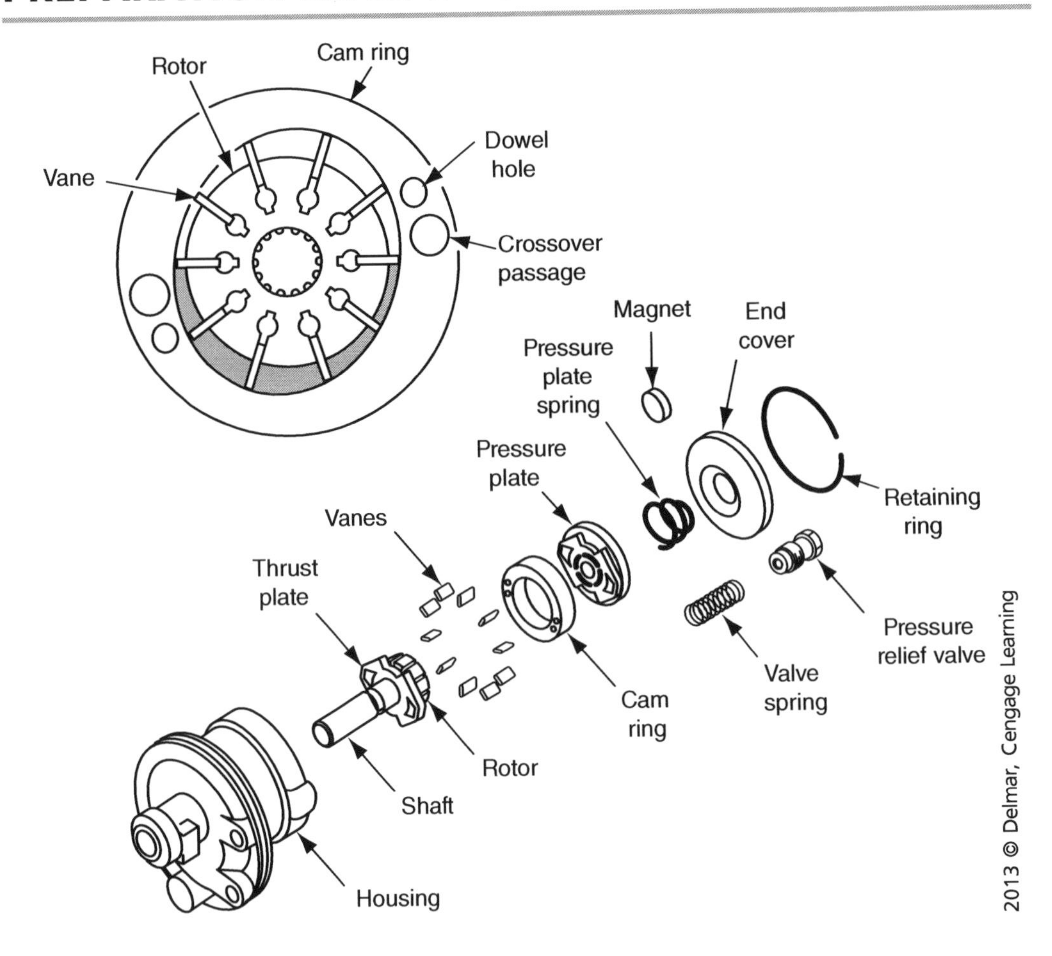

TASK A.8

1. Referring to the figure above, a vane-type power steering pump is being tested. When using a power steering system analyzer with the engine running, the pressure gauge reads low system pressure with the shutoff (load) valve closed. Technician A says the pressure relief valve may be frozen open. Technician B says the pump vanes may be sticking in their slots. Who is correct?

   A. A only
   B. B only
   C. Both A and B
   D. Neither A nor B

   Answer A is incorrect. Technician B is also correct.

   Answer B is incorrect. Technician A is also correct.

   Answer C is correct. Both Technicians are correct. When you test the power steering pump and the pressure gauge or flow meter reads low with the load valve closed, the pressure relief valve may be frozen, or the pump is defective. When the pump vanes stick in their slots the pressure would be low with the shutoff valve closed or open. The power steering pump test procedure varies depending on the type of system. Always follow the procedure in the truck manufacturer's service manual.

   Answer D is incorrect. Both Technicians are correct.

2. The most common type of power steering hydraulic pump used on most Class 8 trucks/ tractors equipped with power steering systems is:

TASK A.11

   A. Gear.
   B. Vane.
   C. Plunger.
   D. Centrifugal.

Answer A is incorrect. Gear pumps are most common on large, off-highway applications.

Answer B is correct. The two most common types are roller pumps and vane pumps.

Answer C is incorrect. Plunger is not a type of power steering pump used in truck/tractor applications.

Answer D is incorrect. Centrifugal is not a type of power steering pump used in truck/tractor applications.

3. Technician A says to inflate mounted tires in a safety cage or using a portable lock ring guard. Technician B says to first mount the tire on the truck, then inflate to the proper tire pressure. Who is right?

TASK D.9

   A. A only
   B. B only
   C. Both A and B
   D. Neither A nor B

Answer A is correct. Only Technician A is correct. You must inflate mounted tires in a safety cage or using a portable lock ring guard.

Answer B is incorrect. You do not mount the tire on the truck before inflation because an improperly fitted rim ring could dislodge, causing injury.

Answer C is incorrect. Only Technician A is correct.

Answer D is incorrect. Technician A is correct.

4. While performing an inspection on a heavy-duty tractor, the technician finds the frame rail buckled. Which of the following might be the cause for this condition?

TASK B.13

   A. Too many holes drilled in the frame
   B. Extreme operating conditions
   C. The wrong bolts used in a repair
   D. A vehicle collision

Answer A is incorrect. Drilling many holes in a frame will weaken it and allow sag to take place.

Answer B is incorrect. Extreme operating conditions such as overloading can also cause sag.

Answer C is incorrect. The wrong bolts used in a frame repair can cause sag due to a release of the reinforcement strength of the repair.

Answer D is correct. A collision can produce the end forces on a frame that would be required to cause a buckling upward of the frame.

TASK C.2

5. Camber may be defined as:

   A. The forward or rearward tilt of the kingpin at the top.
   B. The tracking angle of the tires from a true straight-ahead track.
   C. The inward or outward tilt of the top of the wheel when viewed from the front of the vehicle.
   D. The amount in degrees that the top of the kingpin inclines away from vertical as viewed from the front of the vehicle.

   Answer A is incorrect. The forward or rearward tilt of the kingpin at the top is caster.

   Answer B is incorrect. The tracking angle of the tires from a true straight-ahead track is toe.

   Answer C is correct. Camber is the inward or outward tilt of the top of the wheel when viewed from the front of the vehicle.

   Answer D is incorrect. The amount in degrees that the top of the king-pin inclines away from vertical as viewed from the front of the vehicle is KPI.

TASK A.12

6. A technician finds that the power steering system is overheating. This condition may be caused by all of the following EXCEPT:

   A. A loose steering shaft flex joint.
   B. Underlubricated ball joints.
   C. A kink or pinch in the fluid return line.
   D. Blocked airflow across the heat exchanger.

   Answer A is correct. A steering shaft flex joint is the flexible connector joint between the steering shaft and the steering gear. Being loose, the flex joint would not cause the power steering to overheat.

   Answer B is incorrect. Underlubricated ball joints in the steering joints could make the wheels hard to turn, causing the power steering to work harder and overheat.

   Answer C is incorrect. A blocked or restricted fluid return line would hinder the fluid flow to the power steering heat exchanger, reducing its ability to dissipate the heat from the fluid.

   Answer D is incorrect. Blocked airflow across the heat exchanger would not allow dissipation of heat from the fluid.

TASK C.5

7. Suspension components are replaced on a medium-duty truck with single-air-valve rear air suspension. Technician A says a tape measure can be used to measure from a frame straightedge to the drive axle. Technician B says if the distance from the axle to a fixed straightedge is different on each side of the suspension, one of the front equalizing beam bushings may be worn. Who is correct?

   A. A only
   B. B only
   C. Both A and B
   D. Neither A nor B

   Answer A is incorrect. A tape measure cannot be used to measure from a frame straightedge to the drive axle. A trammel bar should be used.

   Answer B is correct. Only Technician B is correct. If the distance from one side of the axle to a fixed straightedge is different compared to the other side, one of the front equalizing beam bushings may be worn. Worn bushings allow the axle to shift fore and aft and not run parallel to the frame rail.

   Answer C is incorrect. Only Technician B is correct.

   Answer D is incorrect. Technician B is correct.

8. Technician A says that the minimum and maximum air pressure going to the air bags is controlled by the pressure protection valve. Technician B says that the minimum air pressure in the bags should not to be less than 3 psi. Who is correct?

TASK B.8

   A. A only
   B. B only
   C. Both A and B
   D. Neither A nor B

   Answer A is incorrect. The minimum and maximum air pressure going to the bags is controlled by the hand control valve, not a pressure protection valve.

   Answer B is correct. Damage to suspension components may occur if the air pressure is below 3 psi.

   Answer C is incorrect. Only Technician B is correct.

   Answer D is incorrect. Technician B is correct.

9. Excessive tire wear is noted on a heavy-duty tractor. Which of the following suspension components is the LEAST LIKELY cause?

TASK D.1

   A. Bent spindle
   B. Improper tie rod setting
   C. Center cross tube bushing
   D. A preloaded axle wheel bearing

   Answer A is incorrect. A bent spindle would cause inside or outside tread wear.

   Answer B is incorrect. An improper tie rod setting will affect toe, causing tire wear.

   Answer C is incorrect. A defective center cross tube bushing can allow for play and varying toe. Toe and camber are both high tire wear angles.

   Answer D is correct. Preloaded axle wheel bearings would be LEAST LIKELY to cause excessive tire wear. This could, however, cause premature bearing failure.

10. When discussing poppet relief valve adjustment, a technician turns the steering wheel to the left until the steering effort increases. He then measures the distance between the left stop screw and the axle stop. The clearance is found to be 1/4 inch. After checking the stop screw adjustment, the technician finds that the stop screw is within adjustment specifications. The technician should:

TASK A.15

   A. Adjust the stop screw to maintain a 1/8-inch clearance.
   B. Adjust the stop screw to a clearance of 3/8 inch.
   C. Adjust the poppet valves until the clearance is 1/8 inch.
   D. Do nothing. The clearance is okay.

   Answer A is incorrect. If the clearance between the stop screw and axle stop is not within spec, and the stop screw adjustment is adjusted properly, an adjustment to the poppet valve is necessary. This would set the proper travel of the piston within the steering gear and allow for the pressure relief valves to function properly.

   Answer B is incorrect. If the stop screw is adjusted to reduce the clearance between the stop screw and the axle stop, this might reduce the turning radius of the steering system.

   Answer C is correct. The poppet valves should be adjusted to decrease the clearance to 1/8-to 3/16-inch clearance. This sets the proper travel of the piston and relief valves within the steering gear.

   Answer D is incorrect. The poppet valves are out of adjustment and need to be adjusted for proper operation. If left alone, damage to the steering components might result.

TASK B.5

11. When performing an axle alignment on a spring suspension with torque rods, all of the following apply EXCEPT:

    A. The adjustment is made through a lower adjustable torque rod.
    B. The shims are used between the torque rod front and spring hanger bracket.
    C. The adjustment is made with an eccentric bushing at the torque leaf.
    D. The adjustment is made through the upper adjustable torque rods.

    Answer A is incorrect. Axle alignment adjustment on a spring suspension with torque rods can be made through a lower adjustable torque rod.

    Answer B is incorrect. Shims can be used between the torque rod front and spring hanger bracket to align the rear axle.

    Answer C is correct. Some suspensions utilize a torque leaf in place of a torque rod. Adjustment can be made with an eccentric bushing at the torque leaf. However, the question is referring to how a torque arm may be adjusted, not a torque leaf.

    Answer D is incorrect. Adjustments made through the upper torque arms generally set the axle position from side to side and also pitch may be used to position the axle housing to establish the proper driveline angle. This would be part of an axle alignment.

TASK A.1

12. Two technicians are talking about inspection of collapsible steering columns. Technician A says that a dial indicator stem should be placed against the lower end of the steering shaft and then the steering wheel should be rotated. If the runout on the dial indicator exceeds the OEM specification, the steering shaft is bent and must be replaced. Technician B says that if the steering shaft is not bent but shows sheared injection plastic, the shaft must be replaced. Who is correct?

    A. A only
    B. B only
    C. Both A and B
    D. Neither A nor B

    Answer A is incorrect. Technician A is also correct.

    Answer B is incorrect. Technician B is also correct.

    Answer C is correct. Both Technicians are correct. If the runout on the dial indicator exceeds the OEM specification, the steering shaft is bent and must be replaced. Also, if the steering shaft is not bent but shows sheared injection plastic, the shaft must be replaced.

    Answer D is incorrect. Both Technicians are correct.

13. A driver complains that the front end hops, and upon inspection a technician notes that there are cupping marks around the inner circumference of the tire. Which of these is the LEAST LIKELY cause?

TASK D.2

A. Shock absorbers

B. Worn spring shackle pins and bushings

C. Static wheel imbalance

D. Camber angle out of spec

Answer A is incorrect. Worn shock absorbers can cause cupping marks around the tire. Worn shock absorbers allow the tire to bounce causing a cupping mark in the tire. An out-of-specification toe setting causes feathered tire tread wear. Underinflation will wear the edges of the tire. Overinflation will wear the center of the tire. Camber causes wear on one side of the tire. The wear occurs on the inside of the tire with negative camber or the outside of the tire with positive camber.

Answer B is incorrect. Worn spring shackle pins and bushings will cause cupping around the inner circumference of the tire. As with worn shocks, this condition allows the tire to bounce causing a cupping mark in the tire.

Answer C is incorrect. Static imbalance allows the tire to bounce causing a cupping mark in the tire.

Answer D is correct. Camber causes wear on one side of the tire around the entire circumference. The wear occurs on the inside of the tire with negative camber or the outside of the tire with positive camber.

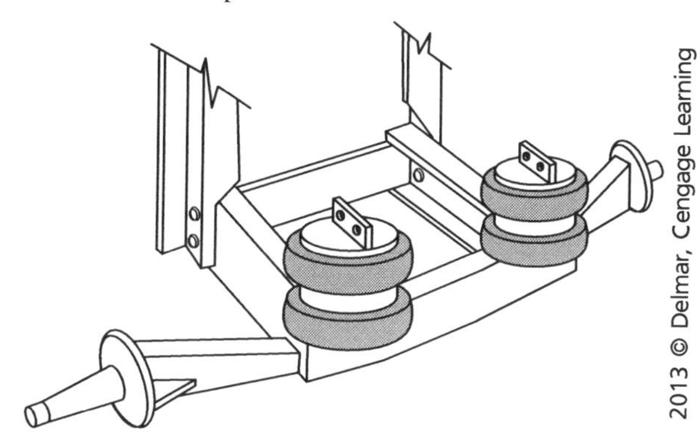

2013 © Delmar, Cengage Learning

14. Referring to the figure above, while discussing the lift axle, Technician A states that the lift controls may be mounted in the cab or mounted externally. Technician B says that when raising or lowering the lift, system air pressure may drop. Who is correct?

TASK B.8

A. A only

B. B only

C. Both A and B

D. Neither A nor B

Answer A is incorrect. Technician B is also correct.

Answer B is incorrect. Technician A is also correct.

Answer C is correct. Both Technicians are correct. Lift axle controls may be mounted internally or externally. When raising or lowering the axle, system air pressure may drop. The axle should be in the full up or down position when the system reaches maximum cut-out pressure.

Answer D is incorrect. Both Technicians are correct.

TASK B.3

15. During an inspection, you see wetness on the shock body. What is your next step?

A. Perform a shock leak test.

B. Replace the wet shock.

C. Replace all shocks.

D. Determine if the shock is actually leaking.

Answer A is incorrect. Before performing a shock leakage test, a visual inspection should be performed to determine if the shock is actually leaking or if the wetness noted is from another source.

Answer B is incorrect. Replacing the shock would be the last step in the process and would only be done after performing a visual inspection and shock leakage test.

Answer C is incorrect. It is not necessary to replace all of the shock absorbers until a visual inspection and shock leakage test are first performed. It is recommended that if one shock is determined to have failed, that all shocks on that axle be replaced to ensure equal dampening across the axle.

Answer D is correct. The next step is to determine if the shock is actually leaking. Wetness on the shock body is normal and does not always indicate that the shock is leaking.

TASK C.6

16. Toe-out on turns is defined by what component?

A. Pitman arm

B. Ackerman arm

C. Drag link

D. Steering control arm

Answer A is incorrect. The pitman arm is attached to the steering gear output shaft and links the steering gear to the drag link. It changes the rotating motion of the steering gear to axial motion.

Answer B is correct. Ackerman arms control the turning radius and allow for each front wheel to rotate on different arcs while turning.

Answer C is incorrect. The drag link connects the pitman arm to the upper steering arm.

Answer D is incorrect. The steering control arm (upper steering arm) is connected to the drag link and is the link between the drag link and the steering knuckle.

TASK B.5

17. Of the following, which is the LEAST LIKELY method used to adjust the axle alignment on a spring suspension with torque rods?

A. Adjustment is made through the lower adjustable torque rod.

B. Shims are used between the torque rod front and spring hanger bracket.

C. Adjustment is made with an eccentric bushing at the torque leaf.

D. Adjustment is made through the upper adjustable torque rod.

Answer A is incorrect. On spring suspensions utilizing torque rods, the fore and aft axle alignment is made by lengthening or shortening the adjustable torque rod.

Answer B is incorrect. On some suspensions with torque arms, both left and right torque arms are of fixed length. In this case, axle alignment adjustments are made by adding or removing shims between the front hanger and the torque arms.

Answer C is correct. Although adjustments may be made by rotating an eccentric bolt on some suspensions, this method is not used with torque arms.

Answer D is incorrect. Lengthening or shortening the upper adjustable torque arm will change axle position when performing an axle alignment.

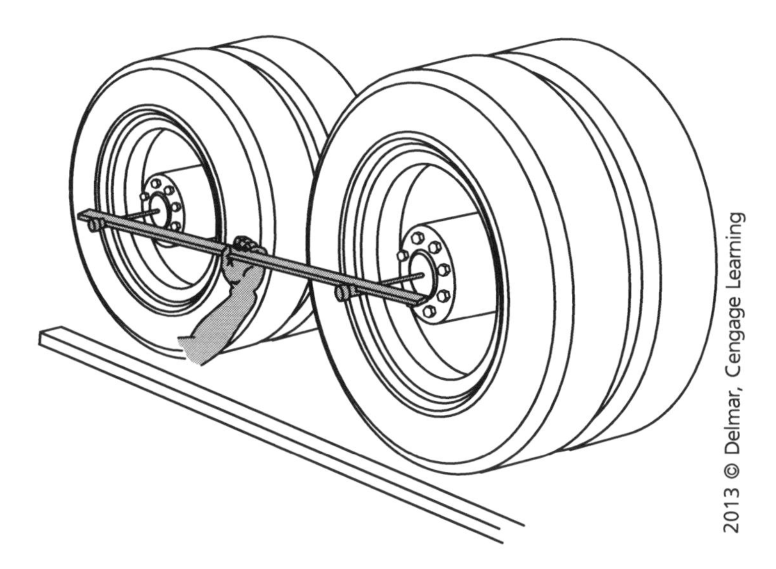

2013 © Delmar, Cengage Learning

18. Referring to the figure above, what is being checked by the technician in the figure?

TASK C.5

   A. Wheel-bearing end-play
   B. Tandem axle spread
   C. Rear toe-in
   D. Rear wheel runout

Answer A is incorrect. The technician is not measuring wheel-bearing end-play. Wheel bearing end-play is checked using a dial indicated on the hub.

Answer B is correct. The technician is measuring tandem axle spread. When the rear axle is properly aligned, the position of the forward rear axle is found with a tram bar. The forward rear axle is then adjusted until it is parallel with the rear axle. The alignment procedures are the same for a trailer, except that an extender target is clamped to the kingpin. The extender target ensures that the rear axle is aligned with the center of the kingpin.

Answer C is incorrect. The technician is not adjusting rear toe-in. Rear axle toe would be checked with a tram bar and tire scribe.

Answer D is incorrect. The technician is not measuring rear wheel runout. Wheel runout is checked by placing a runout gauge against the tire or wheel and rotating the assembly.

19. Which of the following transfers the trailer weight to the fifth wheel?

TASK B.17

   A. The bolster plate
   B. The kingpin
   C. The pintle hook
   D. The drawbar

Answer A is correct. The bolster plate transfers the weight of the trailer to the fifth wheel.

Answer B is incorrect. The kingpin is designed to securely engage with the fifth wheel locking mechanism and to permit articulation.

Answer C is incorrect. The pintle hook is not connected to the fifth wheel.

Answer D is incorrect. The drawbar is not part of the fifth wheel.

TASK A.20

20. Missing, damaged, or out-of-specified adjustment wheel stops may cause:

    A. Wheel imbalance.
    B. Excessive turning radius.
    C. Steering wheel to be off center.
    D. Steering wheel nibble on turns.

    Answer A is incorrect. Wheel imbalance and wheel stops do not affect each other.

    Answer B is correct. When wheel stops are missing, damaged, or out-of-specified adjustment range it can cause excessive turning radius.

    Answer C is incorrect. Wheel stops do not have any effect on the position of the steering wheel.

    Answer D is incorrect. Wheel stops will not cause steering wheel nibble on turns.

TASK C.1

21. Technician A says a badly worn tie rod end can cause steering wander. Technician B says you check tie rod end wear by applying hand pressure as close to the socket as possible. Who is correct?

    A. A only
    B. B only
    C. Both A and B
    D. Neither A nor B

    Answer A is incorrect. Technician B is also correct.

    Answer B is incorrect. Technician A is also correct.

    Answer C is correct. Both Technicians are correct. A badly worn tie rod end can cause steering wander because the spindle and wheel assemblies are not tightly linked together. This allows for some independent movement at each wheel end, which would contribute to steering wander. Check tie rod end wear by applying hand pressure as close to the socket as possible, and pushing and pulling in a vertical motion.

    Answer D is incorrect. Both Technicians are correct.

TASK A.6

22. While inspecting the power steering system on a tractor, the fluid is found to be discolored. All of the following could cause this EXCEPT:

    A. The wrong type of fluid.
    B. Mixed brands of the recommended fluid.
    C. Water mixed with fluid.
    D. An overheated condition.

    Answer A is incorrect. An incompatible fluid will contaminate and discolor the power steering fluid.

    Answer B is correct. Mixed brands of fluid are compatible. Dye is added to different fluid types for identification. Mixing different brands will not discolor existing fluids if the same type is used.

    Answer C is incorrect. Water in power steering fluid will cause foaming and a milky look to the fluid.

    Answer D is incorrect. Overheated fluid will turn brown, and it will have a burned smell.

23. When using a trammel gauge to perform an axle alignment on a vehicle equipped with a tandem axle, Technician A states that the front drive axle should be aligned to the frame. Technician B says that the rear drive axle should be aligned to the frame. Who is correct?

TASK B.5

A. A only
B. B only
C. Both A and B
D. Neither A nor B

Answer A is incorrect. The front drive axle should be aligned to the rear drive axle, using the trammel gauge to measure the distance from the center of the rear drive axle hub to the center of the front drive axle hub. The front axle is not aligned to the frame.

Answer B is correct. Only Technician B is correct. The rear axle is aligned to the frame using the trammel gauge and plumb bob and then the front drive axle is aligned to the rear drive axle.

Answer C is incorrect. Only Technician B is correct.

Answer D is incorrect. Technician B is correct.

24. During a preventive maintenance inspection, the power steering pump mounting bracket bolts are found to be loose. Technician A says elongated mounting holes in the power steering pump bracket may cause a noise while in operation. Technician B says worn holes in the power steering pump mounting bracket could cause premature belt wear. Who is correct?

TASK A.11

A. A only
B. B only
C. Both A and B
D. Neither A nor B

Answer A is incorrect. Technician B is also correct.

Answer B is incorrect. Technician A is also correct.

Answer C is correct. Both Technicians are correct. Elongated bolt holes are usually signs of movement. This would allow belt misalignment and a lack of belt tension, which could cause noisy operation and premature belt wear.

Answer D is incorrect. Both Technicians are correct.

25. What is the LEAST LIKELY purpose for using a tire restraining device?

TASK D.9

A. To protect the technician during tire inflation
B. To inflate a tire that has been damaged to check for leaks
C. To contain the tire and rim parts in the event of an explosion
D. To hold the tire upright in order to measure its circumference

Answer A is incorrect. Using a tire restraining device would help protect the technician in the event of a tire wheel separation.

Answer B is incorrect. Tires should be aired up in a restraining device to constrain all rim wheel components in the event of an explosive separation of a multi-piece rim wheel or during the sudden release of the contained air of a single-piece rim wheel.

Answer C is incorrect. The purpose of the restraining device is to constrain all rim wheel components during an explosive separation of a multi-piece rim wheel or during the sudden release of the contained air of a single-piece rim wheel.

Answer D is correct. Measuring a tire's circumference is done outside of the tire restraining device once the tire is aired up to the proper inflation pressure.

TASK C.1

26. All of the following can cause a front steer axle pulling condition EXCEPT:
    A. A dragging brake.
    B. An out-of-adjustment brake.
    C. Incorrect brake timing.
    D. An incorrect crack pressure relay valve.

Answer A is incorrect. A dragging brake will affect the rolling resistance of the tire/wheel assembly and might cause a front steer axle pull.

Answer B is incorrect. Brakes out of adjustment would be a cause for a pull condition on a steer axle, especially when braking. If both brakes are not adjusted equally, then an application imbalance would occur contributing to the pull condition.

Answer C is incorrect. Brake timing is affected by both pneumatic and mechanical factors. If the brake timing is not equal across the axle, one brake will apply before the other and could cause a steer axle pull condition.

Answer D is correct. Incorrect crack pressure would not cause a pull condition since it would affect both left and right brake assemblies equally.

TASK B.14

27. All of the following are functions of the cross-members of a chassis frame EXCEPT:
    A. To control axial rotation and longitudinal motion of the rails.
    B. To protect wires and tubing that are routed from one side of the vehicle to the other.
    C. To reduce torsional stress transmitted from one rail to the other.
    D. To provide a mounting surface for the fifth wheel assembly.

Answer A is incorrect. Controlling axial rotation and longitudinal motion of the rails is a function of the cross-members.

Answer B is incorrect. Protecting wires and tubing that are routed from one side of the vehicle to the other is a function of the cross-members.

Answer C is incorrect. Reducing torsional stress transmitted from one rail to the other is a function of the cross-members.

Answer D is correct. The fifth wheel mounts to the frame rails, not the cross-members.

TASK C.4

28. When adjusting the toe setting on a truck equipped with radial steering tires, which of the following specifications would apply?
    A. 1/16 inch
    B. 1/8 inch
    C. 3/16 inch
    D. 1/4 inch

Answer A is correct. A toe setting of 1/16 inch allows the tires to assume a neutral toe position when traveling down the road.

Answer B is incorrect. A setting of 1/8 inch will cause the tires to remain slightly toed in. Tire wear will occur with this setting.

Answer C is incorrect. A setting of 3/16 inch will cause the tires to remain toed in. Tire wear will occur with this setting.

Answer D is incorrect. A setting of 1/4 inch will cause the tires to remain toed in. Tire wear will occur with this setting.

29. When adjusting vehicle ride height, which of the following applies?

A. The tractor suspension must be in a laden position with a full load applied.
B. The air system pressure must be at 100 psi.
C. Use safety stands with a sufficient load rating to support the vehicle.
D. Move the height control valve in an upward motion to deflate the air bags.

Answer A is incorrect. The suspension should be in a laden position, but not loaded.

Answer B is correct. Most manufacturers suggest that a minimum of 100 psi is necessary.

Answer C is incorrect. Ride height is checked with the vehicle on the ground.

Answer D is incorrect. Moving the valve lever upward will increase air pressure to the bags.

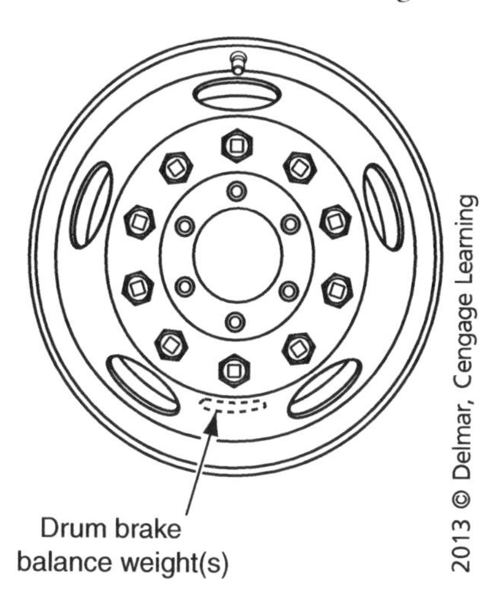

30. Referring to the figure above, a technician is performing a static balance procedure on a truck wheel. Technician A says the maximum wheel weight per tire should not exceed 18 ounces. Technician B says if 16 ounces of weight is required at one spot, use an 8-ounce weight on each side of the rim directly across from each other. Who is correct?

A. A only
B. B only
C. Both A and B
D. Neither A nor B

Answer A is incorrect. Technician B is also correct.

Answer B is incorrect. Technician A is also correct.

Answer C is correct. Both Technicians are correct. The maximum allowable additional wheel weight per tire should not exceed 18 ounces. If more weight is required, it is suggested that the tire be removed from the rim/wheel assembly, rotated 180 degrees and remounted. This will, in many cases, bring the assembly within the acceptable limits. When an out-of-balance weight of 16 ounces is located in one spot, 8 ounces should be placed directly across from each other on either side of the tire. This ensures that the dynamic (side-to-side) balance is not affected.

Answer D is incorrect. Both Technicians are correct.

TASK C.1

31. A vehicle suddenly veers to the right or left after striking a bump with the front wheels. Which of these is the LEAST LIKELY cause?

A. A loose idler arm
B. A damaged relay rod
C. A worn tie rod end
D. A wheel that is out of balance

Answer A is incorrect. If the idler arm is worn and loose, the integrity of the steering linkage is compromised. The looseness would allow the tire and wheel assembly to move when hitting a bump and could cause the vehicle to veer.

Answer B is incorrect. With this condition, the vehicle might suddenly veer to the left or right after hitting a bump because a damaged relay rod might change the toe position of the tires causing steering instability.

Answer C is incorrect. Worn tie rod ends would not keep the steer tires running parallel to each other while traveling straight ahead. This might cause the vehicle to veer after hitting a bump.

Answer D is correct. An out-of-balance wheel may cause a shimmy or vibration, but it would not cause the vehicle to veer after hitting a bump.

TASK C.1

32. After the front springs were replaced on his tractor, a driver complains that it requires excessive steering effort while turning and that the steering wheel return is too fast. Which of these is the most likely cause?

A. Incorrect turning angle
B. Too much negative camber
C. Too much positive caster
D. Underinflated tires

Answer A is incorrect. An incorrect turning angle might cause tire scrub, but will not cause excessive steering effort and cause the steering wheel to return too fast.

Answer B is incorrect. Too much negative camber might cause inside tire tread wear, but will not cause oversteer and cause the steering wheel to return too fast.

Answer C is correct. The most likely cause is excessive positive caster. If the castor setting is too positive, the weight of the vehicle being projected through the kingpin is too far out in front of the vehicle. When this occurs, the tire/wheels want to travel straight ahead and it takes an increased steering effort to turn. Once the turn is completed, because the tire/wheels want to travel straight ahead, the steering wheel will return to center violently.

Answer D is incorrect. Underinflated tires may cause higher steering effort but will not cause the steering wheel to return fast.

33. Technician A says that the pitman arm converts output torque from the steering gear into the control force applied to the drag link. Technician B says that when removing the pitman arm, scribe both the pitman arm and the steering gear shaft prior to removal. Who is correct?

TASK A.16

A. A only
B. B only
C. Both A and B
D. Neither A nor B

Answer A is incorrect. Technician B is also correct.

Answer B is incorrect. Technician A is also correct.

Answer C is correct. Both Technicians are correct. The purpose of the pitman arm is to convert rotational torque to lateral force. Before removing the pitman arm, scribe the arm and shaft so that they may be aligned when reinstalled.

Answer D is incorrect. Both Technicians are correct.

34. All of the following are a result of loose U-bolts EXCEPT:

TASK B.11

A. Axle seat damage.
B. A broken center bolt.
C. Leaf spring breakage between the U-bolts.
D. Leaf spring breakage outside of the U-bolts.

Answer A is incorrect. Axle seat damage can be caused by loose U-bolts.

Answer B is incorrect. A broken center bolt can be caused by loose U-bolts.

Answer C is incorrect. Leaf spring breakage between the U-bolts can be caused by loose U-bolts.

Answer D is correct. Loose U-bolts will not cause leaf spring breakage outside of the U-bolts. Spring breakage outside of the U-bolts may be caused by overload or fatigue.

35. The two main designs of truck wheels are:

TASK D.3

A. Single piece and split side rims.
B. Aluminum and hub-piloted.
C. Disc and drum.
D. Hub and spoke.

Answer A is correct. The two main designs of truck wheels are single piece and split side rims.

Answer B is incorrect. Aluminum is a material used in some single-piece wheels, not a wheel design, and hub-piloted is a design feature of some hubs where wheel pilot pads are machined into the hub to center the wheel.

Answer C is incorrect. Disc and drum are brake component designs.

Answer D is incorrect. Hub and spoke designs are components on split rims.

TASK B.4

36. While discussing leaf spring packs, Technician A states that because leaf springs are clamped together with some force, any movement of the assembly must first overcome friction between the leaves. Technician B states that this condition is called interleaf friction. Who is correct?

 A. A only
 B. B only
 C. Both A and B
 D. Neither A nor B

Answer A is incorrect. Technician B is also correct.

Answer B is incorrect. Technician A is also correct.

Answer C is correct. Both Technicians are correct. As the load is applied to the spring pack, it begins to deflect. This deflection causes bending of each individual leaf in the pack. Because leaf springs are clamped together with some force, any movement of the assembly must first overcome friction between the leaves. This condition is called interleaf friction.

Answer D is incorrect. Both Technicians are correct.

TASK B.6

37. All of the following are true statements regarding walking beams EXCEPT:

 A. Beams can be constructed of aluminum.
 B. Beams can be constructed of cast steel.
 C. Beams can be constructed of nodular iron.
 D. All manufactured beams can be stressed in either direction.

Answer A is incorrect. Beams can be constructed of aluminum.

Answer B is incorrect. Beams can be constructed of cast steel.

Answer C is incorrect. Beams can be constructed of nodular iron.

Answer D is correct. Aluminum and cast steel beams are manufactured in such a way that they can be installed in any direction. Nodular iron beams may only be stressed in one direction and must be installed with the correct side up.

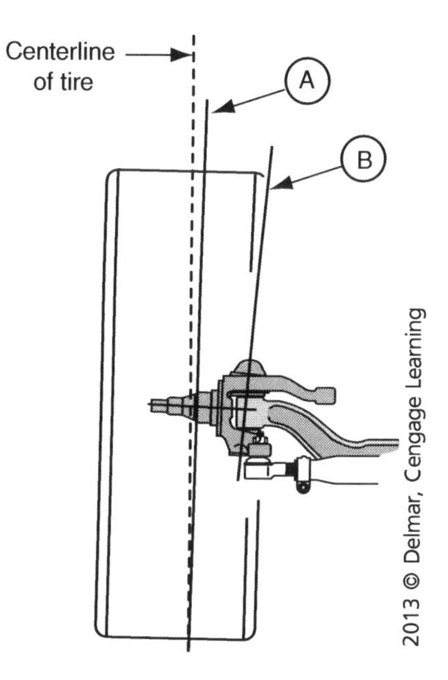

38. Referring to the figure above, what angle is being represented at Letter A?

A. Vertical angle
B. Kingpin inclination
C. Camber
D. Caster

Answer A is incorrect. There is no such angle when talking about front end alignment angles.

Answer B is incorrect. Kingpin inclination is the angle formed between a true vertical line and a line drawn through the kingpin axis.

Answer C is correct. Camber is the angle formed between a true vertical line and a line drawn through the centerline of the tire.

Answer D is incorrect. Castor is the forward or rearward tilt of the kingpin at the top and is not displayed in the figure.

39. After removing the tire from the rim, the technician notices that the rim is cracked. The technician should:

A. Carefully weld the cracked wheel to repair it.
B. Use a brazing rod to rework the wheel and return it for use.
C. Discard it in a pile of other rims.
D. Mark the wheel as unserviceable and remove it from the area.

Answer A is incorrect. Never make an attempt to repair a cracked wheel. Cracked wheels are not to be welded. Welding may distort the wheel and weaken the material surrounding the weld.

Answer B is incorrect. The use of a brazing rod or other welding materials is not recommended. The heat generated through the welding process will change the metal characteristics and the integrity of the wheel assembly will be compromised.

Answer C is incorrect. The wheel must be marked as unserviceable and removed from the area to prevent accidental use.

Answer D is correct. Mark the wheel as unserviceable and remove it from the area. This is an OSHA requirement.

TASK A.6

40. When checking the power steering fluid level, most original equipment manufacturers (OEMs) recommend not checking the fluid level until the system reaches an operating temperature of:

A. 100°F.
B. 240°F.
C. 90°F.
D. 175°F.

Answer A is incorrect. This temperature is too low. It is important to check the level of the power steering fluid at normal operating temperature as fluid expands with heat; checking at a temperature that is too low will not give accurate levels.

Answer B is incorrect. A temperature of 240 degrees is way above normal operating temperature and would be an indicator of a power steering flow issue.

Answer C is incorrect. This temperature is too low. Because fluid expands when heated, it is important to check the fluid level at normal operating temperature to prevent overfilling of the fluid reservoir.

Answer D is correct. OEMs recommend checking the power steering fluid level at a working temperature of at least 175°F.

TASK B.1

41. Technician A says that a machinist protractor may be used to check the caster angle. Technician B says that sometimes, when checking for a twist in the axle, the U-bolts will need to be loosened to relieve axle tension. Who is correct?

A. A only
B. B only
C. Both A and B
D. Neither A nor B

Answer A is incorrect. Technician B is also correct.

Answer B is incorrect. Technician A is also correct.

Answer C is correct. Both Technicians are correct. A machinist protractor may be placed on top of the kingpin cap to measure caster. Also, the U-bolts may need to be loosened to relieve axle tension and place the axle in its natural and unloaded state.

Answer D is incorrect. Both Technicians are correct.

42. You are working with a four-piece nut system with PreSet hub assemblies and the inner nut is torqued to 300 ft-lbs. If the lock ring or spindle washer does not line up with the dowel on the inner nut you should:

TASK D.11

A. Back the inner nut off until the lock ring or spindle washer hole lines up with the dowel.

B. Back the inner nut off to 250 ft-lbs and install the lock ring or spindle washer.

C. Advance the inner nut until the lock ring or spindle nut hole lines up with the dowel.

D. Back the inner nut off to 50 ft-lbs and install the lock ring or spindle washer.

Answer A is incorrect. Backing off the inner nut would increase the end-play and might place the hub adjustment out of specification tolerances.

Answer B is incorrect. Backing off the inner nut to 250 ft-lbs would not guarantee that the lock ring or spindle washer will line up and decreases the clamping force needed to secure the hub to the spindle.

Answer C is correct. Advancing the nut just enough to line up the spindle nut hole with the dowel ensures that the proper end-play is maintained and the proper force is maintained to retain the hub on the spindle.

Answer D is incorrect. Backing off the inner nut to 50 ft-lbs would increase the end-play and would not provide enough force to maintain the hub on the spindle.

43. Of the following, which would be the LEAST LIKELY source of leaking power steering fluid?

TASK A.5

A. Lower sector shaft seal

B. Submersed style pump-to-reservoir surface

C. Supply line double-flare fitting

D. Steering gear input shaft

Answer A is incorrect. Lower sector shaft seals are a likely source of leakage due to seal or shaft wear.

Answer B is correct. If the pump to reservoir surface is submersed, this would not be a likely source of a power steering leak.

Answer C is incorrect. Loose fittings or worn fitting seats would be a good source of a power steering fluid leak.

Answer D is incorrect. Worn input shaft or seal wear would allow for a fluid leak.

44. After a front alignment has been performed, a driver complains of a pull. Technician A says that the tractor will pull to the side with the most positive caster angle setting. Technician B says that the tractor will pull to the side with the most positive camber setting. Who is correct?

TASK C.1

A. A only

B. B only

C. Both A and B

D. Neither A nor B

Answer A is incorrect. A tractor will pull to the side with the most negative castor.

Answer B is correct. Only Technician B is correct. A truck will pull to the side that has the most positive camber setting. Positive camber produces an outward tilt of the wheel at the top, which places more weight at the outer edge of the tire tread. This increases tire rolling resistance, which can cause pull.

Answer C is incorrect. Only Technician B is correct.

Answer D is incorrect. Technician B is correct.

TASK B.14

45. During an inspection, a technician finds a crack in the cross-member web that is extending into the flange. While preparing to repair a frame cross-member, Technician A says to disconnect the truck batteries before welding a frame. Technician B says it is permissible to weld across frame flanges. Who is correct?

A. A only
B. B only
C. Both A and B
D. Neither A nor B

Answer A is correct. Only Technician A is correct. You should disconnect the truck batteries before welding a frame.

Answer B is incorrect. You never weld across frame flanges. The frame strength is weakened and the frame is more susceptible to damage.

Answer C is incorrect. Only Technician A is correct.

Answer D is incorrect. Technician A is correct.

TASK A.13

46. Technician A says noise from the steering gear might be caused by misalignment of the steering column input shaft. Technician B says that noise coming from the manual steering gear assembly of a linkage-assist-type power steering when turning the steering may be caused by low lubricant level. Who is correct?

A. A only
B. B only
C. Both A and B
D. Neither A nor B

Answer A is incorrect. Technician B is also correct.

Answer B is incorrect. Technician A is also correct.

Answer C is correct. Both Technicians are correct. A misaligned steering column input shaft can cause binding noises in the steering gear and a steering gear with low lubricant can produce noise during operation.

Answer D is incorrect. Both Technicians are correct.

TASK C.1

47. A driver complains that while going down the road the vehicle bounces excessively and pulls to the right. Technician A says the first thing that should be done is to replace the front shock absorbers and align the front end. Technician B says the first thing that should be done is to balance the front wheels. Who is correct?

A. A only
B. B only
C. Both A and B
D. Neither A nor B

Answer A is incorrect. No procedures should be performed or components replaced until the vehicle has been road tested and the front end inspected.

Answer B is incorrect. No procedures should be performed or components replaced until the vehicle has been road tested and the front end inspected.

Answer C is incorrect. Neither Technician is correct.

Answer D is correct. Neither Technician is correct. Following a front end complaint no procedures should be performed or components replaced until the vehicle has been road tested and the front end inspected.

48. On a typical single-axle truck system, the Ackerman arms are linked by what component?

TASK C.6

A. A drag link
B. A pitman arm
C. A steering column
D. A tie rod assembly

Answer A is incorrect. The drag link connects the pitman arm to the upper steering arm.

Answer B is incorrect. The pitman arm is attached to the steering gear and links the steering gear to the drag link.

Answer C is incorrect. The steering column is connected to the input shaft of the steering gear.

Answer D is correct. The tie rod assembly connects the Ackerman arms.

49. A truck is dog tracking while being driven down the road. Upon inspection, the technician diagnoses the vehicle and notes that the frame is in a diamond condition. Technician A says that towing another truck with a chain attached to one corner of the frame could be the cause. Technician B says the shifting of the front or rear axles could be the cause. Who is correct?

TASK B.13

A. A only
B. B only
C. Both A and B
D. Neither A nor B

Answer A is correct. Only Technician A is correct. Towing another truck with a chain attached to one corner of the frame will cause a diamond frame condition.

Answer B is incorrect. The shifting of the front or rear axles will not cause frame buckle. The shifting of axles affects alignment and tracking.

Answer C is incorrect. Only Technician A is correct.

Answer D is incorrect. Technician A is correct.

50. The LEAST LIKELY purpose for pre-lubing the spindle and hub cavity during hub installation is:

TASK D.10

A. To pre-charge the wheel bearing with lubricant.
B. To prevent fretting corrosion.
C. To prevent spindle contamination.
D. To aid in hub installation.

Answer A is incorrect. The purpose of pre-charging the hub cavity prior to hub installation is to provide some lubrication for the bearings. This also reduces the amount of oil needed and time when filling the hub after installation.

Answer B is incorrect. The purpose of pre-lubing the spindle prior to hub installation is to prevent fretting.

Answer C is correct. Spindle contamination would occur from outside sources. The spindle and bearings are cleaned prior to installation.

Answer D is incorrect. The purpose of pre-lubing the spindle prior to hub installation is to ensure the bearings and seal do not hang up on the spindle during installation.

PREPARATION EXAM 3 – ANSWER KEY

1. D
2. A
3. D
4. C
5. C
6. C
7. B
8. B
9. C
10. C
11. A
12. A
13. C
14. C
15. B
16. D
17. B
18. B
19. C
20. C
21. B
22. A
23. B
24. B
25. B
26. D
27. A
28. A
29. C
30. A
31. A
32. A
33. D
34. D
35. C
36. C
37. A
38. B
39. D
40. A
41. C
42. A
43. C
44. B
45. B
46. A
47. D
48. D
49. D
50. D

PREPARATION EXAM 3 – EXPLANATIONS

1. Which of these statements is correct when diagnosing truck frame problems?

   A. Tandem-axle tractors have the maximum bending moment occur at the bogie centerline.

   B. Single-axle trucks with van bodies have the maximum bending moment occur just ahead of the rear axle.

   C. Frame buckle may be caused by too many holes drilled in the frame web.

   D. Frame twist may be caused by uneven loading.

TASK B.13

Answer A is incorrect. Tandem-axle tractors have the maximum bending moment occur just ahead of the fifth wheel, not at the bogie centerline.

Answer B is incorrect. Single-axle trucks with van bodies have the maximum bending moment occur just to the rear of the cab, not ahead of the rear axle.

Answer C is incorrect. Frame buckle cannot be caused by too many holes drilled in the frame web. Drilling too many holes causes frame sag.

Answer D is correct. Frame twist may be caused by uneven loading.

2. The driver of a truck complains of a shimmy in the front end. A technician performed a front-end alignment. While taking the unit for a road test, the technician finds that a truck still has a shimmy. Technician A says this could be caused by too much positive caster. Technician B says an improper camber setting can be the cause. Who is correct?

TASK C.1

   A. A only

   B. B only

   C. Both A and B

   D. Neither A nor B

Answer A is correct. Only Technician A is correct. A shimmy after an alignment has been performed can be caused by too much positive caster. Harsh riding may be caused by excessive positive caster because the caster line is actually aimed at some road irregularities. Excessive positive caster may cause front wheel shimmy from side to side at low speeds.

Answer B is incorrect. An improper camber setting might cause the tires to wear but will not cause wheel shimmy.

Answer C is incorrect. Only Technician A is correct.

Answer D is incorrect. Technician A is correct.

TASK C.3

3. When adjusting caster, all of the following statements are true EXCEPT:

   A. Use only one shim on each side.

   B. The width of the shim should equal the width of the spring.

   C. Do not use shims of more than 1 degree difference from side to side.

   D. Reverse the shims from left to right to correct for a twisted axle.

Answer A is incorrect. Only one shim of proper thickness should be used on each side to prevent shim shifting and breakage.

Answer B is incorrect. The shim width should be the same width as the spring width to prevent shifting and ensure proper clamping force is maintained by the U-bolts.

Answer C is incorrect. This might induce the same effect as a twisted axle.

Answer D is correct. Twisted axles should be replaced.

TASK B.17

4. What is one of the standard SAE kingpin sizes used on trailers?

   A. 1.5 inches

   B. 2.5 inches

   C. 3.5 inches

   D. 4.5 inches

Answer A is incorrect. There are two standard kingpin sizes used on the trailer upper coupler assembly. One is 2 inches and the other is 3.5 inches.

Answer B is incorrect. 2.5 inches is not a standard SAE kingpin size. The standard sizes are 2 inches and 3.5 inches.

Answer C is correct. One of the standard SAE kingpin sizes used on trailers is 3.5 inches.

Answer D is incorrect. One of the standard SAE kingpin sizes used on trailers is 3.5 inches. The standard sizes are 2 inches and 3.5 inches. There is not a 4.5-inch kingpin available for trailer couplers.

TASK B.14

5. A frame reinforcement is needed on a tractor. Which of the following is true when using C-channel stock for the repair?

   A. It should be welded directly to the frame rail.

   B. It must be of the same dimensions as the original frame rail.

   C. It must have the same yield strength as the original frame.

   D. It should only be installed on the outside of the original frame rail.

Answer A is incorrect. C-channel reinforcements are not welded to the original frame rail. Welding would weaken the original rail at the welds.

Answer B is incorrect. The C-channel must be smaller than the original rail in order to fit inside the rail.

Answer C is correct. The yield strength must be the same as the original rail. This is due to the flex characteristics of different yield strength materials.

Answer D is incorrect. C-channel reinforcements should fit inside of the original rail.

6. While performing an inspection on a power steering system pump, a growling noise is present and air has been found in the system. Technician A says that if the growling noise is still present after the fluid level is checked and the air is bled from the system, the pump bearings or other components may be damaged. Technician B states that when the power steering pump pressure is lower than specified, pump replacement or repair is indicated. Who is correct?

   A. A only
   B. B only
   C. Both A and B
   D. Neither A nor B

   Answer A is incorrect. Technician B is also correct.

   Answer B is incorrect. Technician A is also correct.

   Answer C is correct. Both Technicians are correct. Growling noises and low pump pressures are indications of internal component damage.

   Answer D is incorrect. Both Technicians are correct.

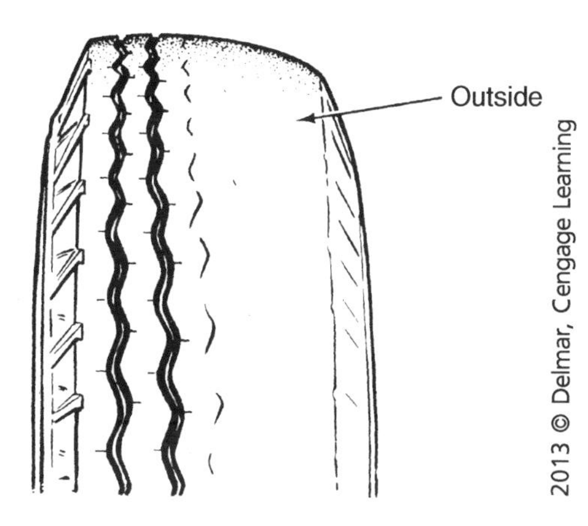

7. Referring to the figure above, what does this tire wear pattern indicate?

   A. Excessive toe-out
   B. Excessive positive camber
   C. Excessive positive caster
   D. Overinflation

   Answer A is incorrect. The wear shown is not featheredging caused by toe-out.

   Answer B is correct. The tire wear shown is outer tread wear that is caused by excessive positive camber.

   Answer C is incorrect. Caster is not considered a tire wear angle; however, excessive positive camber roll will cause more camber roll, which will cause tire wear.

   Answer D is incorrect. Overinflation will wear the center of the tread.

TASK C.2

8. While performing a front wheel alignment, Technician A says that a camber setting of 1 3/4 degree for the curb side and 1 1/4 degree for the driver side is typical. Technician B says that camber settings depend on the amount of KPI that is built into the axle. Who is correct?

A. A only
B. B only
C. Both A and B
D. Neither A nor B

Answer A is incorrect. The preferred settings are of 1 3/4 degree for the driver side and 1 1/4 degree for the curb side.

Answer B is correct. Only Technician B is correct. The amount of camber is determined by how much KPI is built into the axle.

Answer C is incorrect. Only Technician B is correct.

Answer D is incorrect. Technician B is correct.

TASK A.15

9. Technician A says that poppet valves must be adjusted properly to prevent excessive pressure on the steering linkages when the steering wheel is turned fully to the left or right. Technician B says that the poppet valve adjustment is performed with the steering gear installed on the vehicle and the power steering fluid at normal operating temperature. Who is correct?

A. A only
B. B only
C. Both A and B
D. Neither A nor B

Answer A is incorrect. Technician B is also correct.

Answer B is incorrect. Technician A is also correct.

Answer C is correct. Both Technicians are correct. The poppet valves help relieve pressure on the system and components. When testing or adjusting the poppet valves, the steering gear should be installed and the fluid should be at operation temperature.

Answer D is incorrect. Both Technicians are correct.

TASK C.5

10. Technician A says steer axle misalignment may cause abnormal steer axle tire wear. Technician B says drive axle misalignment may cause abnormal steer axle tire wear. Who is correct?

A. A only
B. B only
C. Both A and B
D. Neither A nor B

Answer A is incorrect. Technician B is also correct.

Answer B is incorrect. Technician A is also correct.

Answer C is correct. Both Technicians are correct. Camber, toe, KPI, and turning radius are all tire wearing angles. In addition, if the drive axle is not aligned to the frame rails correctly, the result is tire scrub on the front steer tires. Rear axle misalignment may be noted because one front tire will be worn on the outside edge and the opposite tire will be worn on the inside edge.

Answer D is incorrect. Both Technicians are correct.

11. A driver complains that his tractor/trailer is leaning to the left and air is heard leaking out of the left-rear height control valve. He also states that the air system pressure went to 0 psi. The cause of the low air pressure in the braking system could be:

TASK B.10

A. A defective pressure protection valve.
B. A defective pressure reduction valve.
C. A defective height control valve.
D. An improper ride height adjustment.

Answer A is correct. The pressure protection valve is used on auxiliary systems to prevent total air system pressure loss.

Answer B is incorrect. Pressure reduction valves are not used in the air suspension system.

Answer C is incorrect. Height control valves only control the air bag pressure.

Answer D is incorrect. Ride height would not cause this condition.

12. When a technician is performing a close visual inspection of the power steering belt(s) his inspection may reveal all of the following EXCEPT:

TASK A.10

A. Proper belt tension.
B. Premature wear due to misalignment.
C. Correct orientation of dual belt application.
D. Proper belt seating in the pulley.

Answer A is correct. A visual inspection cannot verify proper belt tension. A belt tension gauge should be used to check belt tension.

Answer B is incorrect. A visual inspection of the power steering belts can reveal belt misalignment.

Answer C is incorrect. A visual inspection of the power steering belts can reveal proper dual belt orientation.

Answer D is incorrect. A visual inspection of the power steering belts can reveal proper belt seating in the pulleys.

13. A driver complains that his heavy-duty truck requires excessive steering effort and that the steering wheel return is too fast. Which of these is the most likely cause?

TASK B.10

A. Incorrect turning angle
B. Too much negative camber
C. Too much positive caster
D. Underinflated tires

Answer A is incorrect. An incorrect turning angle might cause tire scrub and tire wear, but it will not cause excessive steering effort and cause the steering wheel to return too fast.

Answer B is incorrect. Too much negative camber will cause decreased effort going into a turn and the wheels will not want to return to center without assistance from the driver, but it will not cause oversteer and cause the steering wheel to return too fast.

Answer C is correct. The most likely cause is excessive positive caster, which projects the vehicle load through the kingpin too far out in front of the axle. This results in the need for increased effort going into the turn. In addition, one of the functions of positive castor is to help bring the wheels back to a straight-ahead position after turning. This is why the steering wheel returns much too fast.

Answer D is incorrect. Underinflated tires may cause higher steering effort but will not cause the steering wheel to return fast.

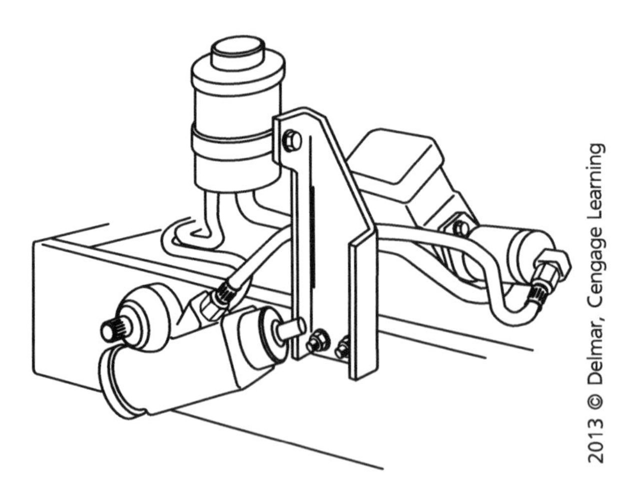
2013 © Delmar, Cengage Learning

TASK A.11

14. Referring to the figure above, the power steering pump is being removed. All of the tasks listed below are required to be performed during removal and replacement EXCEPT:

A. Check the pump mounting holes for wear.

B. Remove the hoses from the pump and cap the fittings.

C. Reuse the o-ring if it is in good condition.

D. Bleed.

Answer A is incorrect. You do check the pump mounting holes for wear.

Answer B is incorrect. You do remove the hoses from the pump and cap the fittings.

Answer C is correct. You never reuse the o-ring even if it is in good condition.

Answer D is incorrect. You bleed the air from the power steering system.

TASK C.4

15. Toe may be defined as:

A. The forward or rearward tilt of the kingpin at the top.

B. The tracking angle of the tires from a true straight-ahead track.

C. The inward or outward tilt of the top of the wheel when viewed from the front of the vehicle.

D. The amount in degrees that the top of the kingpin inclines away from vertical as viewed from the front of the vehicle.

Answer A is incorrect. The forward or rearward tilt of the kingpin at the top is the definition of caster.

Answer B is correct. The tracking angle of the tires from a true straight-ahead track is toe. It is the angle formed by two horizontal lines through the planes of two wheels as viewed from the top of the wheels.

Answer C is incorrect. Camber is the inward or outward tilt of the top of the wheel when viewed from the front of the vehicle.

Answer D is incorrect. KPI (kingpin inclination) is the amount in degrees that the top of the kingpin inclines away from a true vertical line drawn through the center of the tire and wheel assembly as viewed from the front of the vehicle.

16. When installing cups in a cast-iron hub, the hub needs to be heated to between:

TASK D.10

A. 200°F and 220°F.

B. 450°F and 500°F.

C. 180°F and 200°F.

D. Cast-iron hubs are not heated.

Answer A is incorrect. Aluminum hubs are heated to install the bearing cups, not steel hubs.

Answer B is incorrect Steel hubs are not heated to install bearing cups. In addition, 450°F to 500°F is excessive. Excessive heat may distort the component.

Answer C is incorrect. 180°F to 200°F is the temperature specification for installing bearing cups in an aluminum hub. Steel hubs are not heated.

Answer D is correct. Cast-iron hubs are not heated to install the bearing cup. They are generally removed using a brass drift punch and hammer.

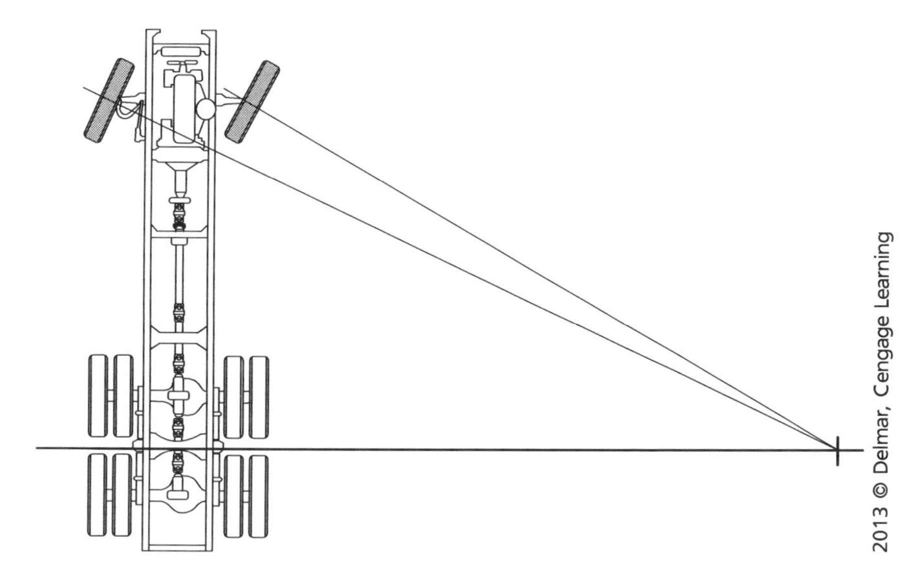

17. Referring to the figure above, Technician A says that most steering knuckles use adjustable wheel stops for setting toe-out on turns. Technician B says that during a turn the front wheels must be turned at different angles to prevent tire scuffing. Who is correct?

TASK C.6

A. A only

B. B only

C. Both A and B

D. Neither A nor B

Answer A is incorrect. Adjustable wheel stops only set the maximum turning angle of the front wheels.

Answer B is correct. Only Technician B is correct. The front wheels are turned at different angles so the wheels can follow separate wheel paths due to the different distances from the center of the turn.

Answer C is incorrect. Only Technician B is correct.

Answer D is incorrect. Technician B is correct.

TASK A.17

18. The drag link connects the pitman arm to which component on the front non-drive steer axle?

 A. The steering gear

 B. The steering arm

 C. The front axle

 D. The torque arm

Answer A is incorrect. The drag link is not directly connected to the steering gear. The pitman arm is connected to the steering gear and changes rotary motion to linear motion. The drag link is the link that connects the pitman arm to the upper steering arm at the spindle.

Answer B is correct. The pitman arm and steering arm are linked through the drag link.

Answer C is incorrect. The front axle is not connected to the drag link or the pitman arm. The front axle is connected to the front springs via U-bolts.

Answer D is incorrect. Torque arms are not used on front non-drive steer axles. Torque arms are used on drive axle assemblies.

TASK B.6

19. Upon inspection, a technician finds broken leaves within the spring pack on a leaf spring-type walking beam suspension. Technician A says that this may be caused by overloading the suspension. Technician B says that the spring pack should be replaced. Who is correct?

 A. A only

 B. B only

 C. Both A and B

 D. Neither A nor B

Answer A is incorrect. Technician B is also correct.

Answer B is incorrect. Technician A is also correct.

Answer C is correct. Both Technicians are correct. Overload is a common cause of leaf spring failure. Also, as a rule, most original equipment manufacturers (OEMs) recommend replacement of the spring pack anytime a leaf is broken. If the spring pack is replaced, it is a good practice to replace the other spring pack on that axle.

Answer D is incorrect. Both Technicians are correct.

TASK D.8

20. Tire matching is being discussed. Technician A says that dual tires are matched to prevent tire tread wear from slippage from uneven surface areas. Technician B says dual drive wheels may be measured using a tire tape measure. Who is correct?

 A. A only

 B. B only

 C. Both A and B

 D. Neither A nor B

Answer A is incorrect. Technician B is also correct.

Answer B is incorrect. Technician A is also correct.

Answer C is correct. Both Technicians are correct. Matching tire sizes on dual wheels prevents tire tread wear from slippage from uneven surface areas. Dual drive wheels may be measured with a tape measure, string gauge, square, tire caliper, or straightedge.

Answer D is incorrect. Both Technicians are correct.

21. Technician A says air spring suspensions use shock absorbers to help maintain axle alignment. Technician B says that shock absorbers are used to control spring oscillations. Who is correct?

TASK B.7

   A. A only
   B. B only
   C. Both A and B
   D. Neither A nor B

   Answer A is incorrect. The purpose of the shock absorbers is to help control air spring oscillation, not axle alignment.

   Answer B is correct. Only Technician B is correct. Shock absorbers are used to control spring oscillation because an air spring does not have any dampening characteristics.

   Answer C is incorrect. Only Technician B is correct.

   Answer D is incorrect. Technician B is correct.

22. Caster may be defined as:

TASK C.3

   A. The forward or rearward tilt of the kingpin at the top.
   B. The tracking angle of the tires from a true straight-ahead track.
   C. The inward or outward tilt of the top of the wheel when viewed from the front of the vehicle.
   D. The amount in degrees that the top of the kingpin inclines away from vertical as viewed from the front of the vehicle.

   Answer A is correct. Caster is the forward or rearward tilt of the kingpin at the top.

   Answer B is incorrect. The tracking angle of the tires from a true straight-ahead track is the definition of toe.

   Answer C is incorrect. The inward or outward tilt of the top of the wheel when viewed from the front of the vehicle is the definition of camber.

   Answer D is incorrect. The amount in degrees that the top of the kingpin inclines away from vertical as viewed from the front of the vehicle is the definition of kingpin inclination.

23. When draining a power steering system, all of the following should be done EXCEPT:

TASK A.7

   A. Drain the fluid by removing the return hose at the remote reservoir fitting.
   B. With the engine stopped, turn the steering wheel fully in each direction.
   C. When the clean fluid begins to discharge from the return hose, shut the engine off.
   D. Perform a final "bleed" on the system once the return hose is reinstalled.

   Answer A is incorrect. When the power steering system is being drained and flushed, the return hose from the gear to the remote reservoir should be disconnected at the remote reservoir to drain the fluid.

   Answer B is correct. The engine should be running when purging the fluid from the gear.

   Answer C is incorrect. The engine should be shut off when clean fluid appears from the return hose.

   Answer D is incorrect. Once the return line has been reinstalled, a final "bleed" of the system should be performed.

TASK D.5

24. A technician finds an irregular wear pattern on a tire that is overinflated. Which wear pattern did the technician find on the tire?

    A. Outside edges of the tire
    B. Center of the tire
    C. Cupping pattern
    D. Inside edges of the tire

Answer A is incorrect. When a tire is underinflated, the vehicle weight is concentrated on the outer edges of the tire, and the center of the tread may not even contact the road surface.

Answer B is correct. Overinflation will wear the center of the tire. If a tire has the specified inflation pressure, the entire tread surface is in contact with the road surface, but overinflation lifts the edges of the tire off the road surface, and the vehicle weight is concentrated on the center of the tread.

Answer C is incorrect. Tire bounce causes a cupping pattern.

Answer D is incorrect. Camber causes wear on the inside of the tire.

TASK D.9

25. When removing a disc wheel from any truck, Technician A says that when a disc wheel needs to be removed from any truck, the right-side wheel will have right-hand threads and the left side left-hand threads. Technician B says to follow good safety practices by wearing safety glasses and not standing in front of a deflating tire. Who is correct?

    A. A only
    B. B only
    C. Both A and B
    D. Neither A nor B

Answer A is incorrect. Not all trucks with disc wheels have right-hand threads on the right and left-hand threads on the left.

Answer B is correct. Only Technician B is correct. Safety glasses should always be worn and positioning your body out of harm's way is a good safety practice when servicing all tire and rim assemblies. Warning: Stay out of the trajectory (danger) zone. Under some circumstances the trajectory may deviate from its standard path.

Answer C is incorrect. Only Technician B is correct.

Answer D is incorrect. Technician B is correct.

TASK B.6

26. Which of the following shop tools would most likely be required to remove old rubber bushings from an equalizing beam-type suspension?

    A. A hammer and chisel
    B. An oxy-acetylene cutting torch
    C. A rosebud
    D. A 50-ton hydraulic press

Answer A is incorrect. Equalizing beam bushings are a press fit design and are installed using a 50-ton hydraulic press. Because of the force needed to install the bushings, the use of a hammer and chisel is not practical. In addition, using a hammer and chisel may damage the bushing bores.

Answer B is incorrect. By using a torch, the technician risks damaging the beam. In addition, once alight, the bushings will burn for a long time producing high heat and noxious fumes.

Answer C is incorrect. Heating the beam with a rosebud will change the temper characteristics of the beam.

Answer D is correct. Bushings should be removed with a hydraulic press.

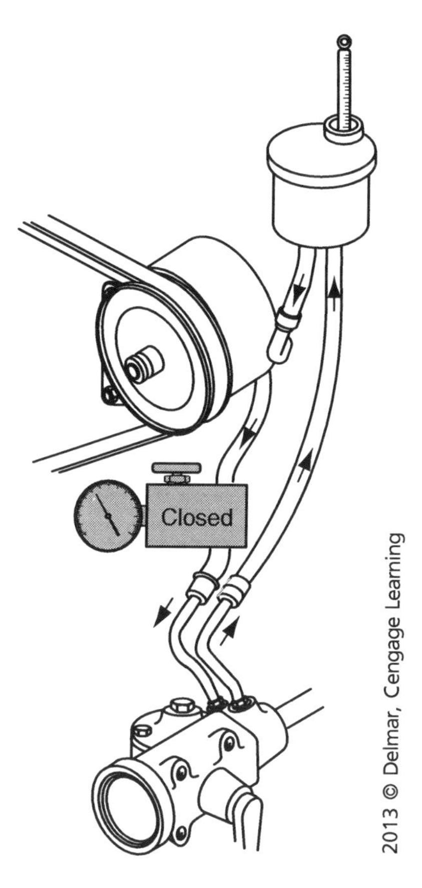

2013 © Delmar, Cengage Learning

27. Referring to the figure above, when using a power steering system analyzer with the engine running, if the flow control valve on the analyzer is closed, what should happen?

A. The system pressure goes up.
B. The system pressure goes down.
C. The wheels turn off-center.
D. The GPM gauge reading increases.

Answer A is correct. The system pressure should increase due to the restriction.

Answer B is incorrect. The system pressure increases, not decreases, when the valve is closed.

Answer C is incorrect. The gear is not being rotated so the wheels will not move.

Answer D is incorrect. The GPM will decrease due to the restriction.

TASK D.3

28. What type of nuts are used to hold the wheels in place on a hub with a hub-piloted mounting system?

A. A flanged nut
B. A nut and washer combination
C. An axle nut
D. A ball-seat nut

Answer A is correct. Flanged nuts are used with hub-piloted mounting systems.

Answer B is incorrect. A nut and washer combination is not a type of wheel retaining components. Flanged nuts are used on hub-piloted wheels and ball-seat nuts are used with stud-piloted wheels.

Answer C is incorrect. Axle nuts retain the axle to the hub, not the wheel assembly.

Answer D is incorrect. Ball-seat nuts are used on stud-piloted wheel systems.

TASK D.7

29. You are balancing the radial tires on a medium truck. Technician A says wheel runout should be measured before the balance procedure. Technician B says after the wheels are balanced, mount the tire with the wheel weights 180 degrees from the brake drum weights. Who is right?

A. A only
B. B only
C. Both A and B
D. Neither A nor B

Answer A is incorrect. Technician B is also correct.

Answer B is incorrect. Technician A is also correct.

Answer C is correct. Both Technicians are correct. Before balancing a radial tire, you mount the tire and check for runout. After the wheels are balanced you should mount the tire with the wheel weights 180 degrees away from the brake drum weights.

Answer D is incorrect. Both Technicians are correct.

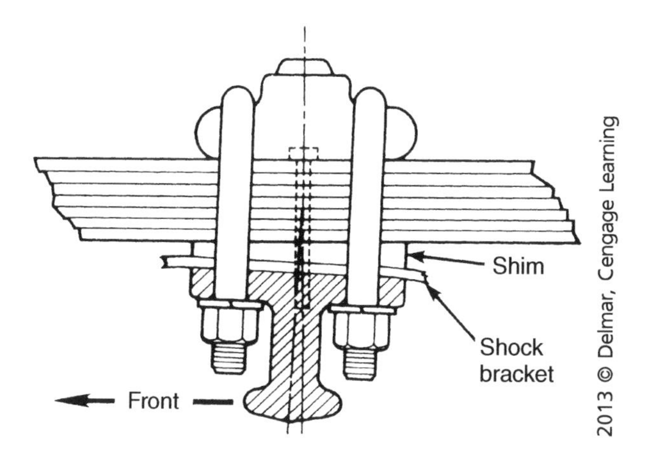

30. Referring to the figure above, a technician has installed the shim to adjust:

TASK C.3

A. Caster.

B. Camber.

C. Toe-in.

D. Spring sag.

Answer A is correct. Caster is typically adjusted by tilting the axle with the use of shims. Caster angle is changed through the use of a caster shim, a metal plate inserted between the springs and the axle pad.

Answer B is incorrect. Camber is not adjustable on a solid axle.

Answer C is incorrect. Toe-in is not adjustable with shims.

Answer D is incorrect. A shim cannot correct spring sag.

31. Two technicians are discussing the replacement procedure for a power steering pump pulley. Technician A states that if there is a retaining nut on the shaft, the pulley is retained by a woodruff key to prevent pulley rotation on the shaft. Technician B says that it is acceptable to use a soft-faced hammer and to tap on the pump shaft to remove the pulley. Who is correct?

TASK A.10

A. A only

B. B only

C. Both A and B

D. Neither A nor B

Answer A correct. Only Technician A is correct. Belt-driven pump pulleys utilizing a retaining nut are held in place by a woodruff key.

Answer B is incorrect. Never use a hammer on the pump shaft. Damage to the shaft or pump may occur.

Answer C is incorrect. Only Technician A is correct.

Answer D is incorrect. Technician A is correct.

TASK B.10

32. A vehicle is found to be "dog tracking." What would be the LEAST LIKELY cause of this condition?

    A. Loose spring shackles
    B. Loose U-bolts
    C. Bent torque arm
    D. Bent frame

Answer A is correct. Loose spring shackles will cause steerability complaints and tire wear but would not cause the vehicle to "dog track."

Answer B is incorrect. Loose U-bolts can break the spring center bolt. If the spring U-bolts are not tightened properly, the axle may shift on the spring. This action may shear off the center bolt. This condition causes improper axle position on the spring, resulting in serious steering problems such as pulling to one side and pulling while braking.

Answer C is incorrect. A bent torque arm would alter the axle position pulling the axle out of square with the vehicle frame. This would position one tire/wheel assembly forward of the other and would cause the vehicle to "dog track."

Answer D is incorrect. A bent frame that has side sway damage would cause the drive axle to be out of alignment. This would cause a "dog track" condition.

TASK D.10

33. Four functions a lubricant provides the wheel bearings are to lubricate, clean the bearings, prevent corrosion, and:

    A. Prevent contamination.
    B. Contain the bearing.
    C. Control the bearing.
    D. Cool the bearing.

Answer A is incorrect. Lubricant does not prevent contamination.

Answer B is incorrect. Hub components contain the bearings.

Answer C is incorrect. Lubricant does not control the bearings.

Answer D is correct. A function of lubrication is to cool the bearing.

TASK C.3

34. A technician is inspecting a front axle assembly. He states that a technician should check for a twisted axle beam if any of the following conditions exist EXCEPT:

    A. The difference in caster angle exceeds 1/2 degree from side to side.
    B. The caster shims in place differ by 1 degree or more.
    C. A low-speed shimmy exists and there is no evidence of looseness elsewhere in the steering system.
    D. Excessive tire wear exists.

Answer A is incorrect. If the difference in caster angle exceeds 1/2 degree from side to side, a twisted axle is indicated.

Answer B is incorrect. If the caster shims in place differ by 1 degree or more, a twisted axle is indicated.

Answer C is incorrect. If a low-speed shimmy exists and there is no evidence of looseness elsewhere in the steering system, a twisted axle is indicated.

Answer D is correct. Caster is a non-tire-wearing angle.

35. Technician A says that the condition of high hitch occurs when the trailer kingpin and bolster plate are positioned too high in the fifth wheel. Technician B says this condition is caused by improper bolster plate position during the coupling process. Who is right?

TASK B.15

A. A only
B. B only
C. Both A and B
D. Neither A nor B

Answer A is incorrect. Technician B is also correct.

Answer B is incorrect. Technician A is also correct.

Answer C is correct. Both Technicians are correct. A high hitch can occur when you position the trailer kingpin (pintle) and bolster plate too high in the fifth wheel. This condition is also caused by improper bolster plate position during the coupling process.

Answer D is incorrect. Both Technicians are correct.

36. Technician A states that axle alignment on a tandem rear axle suspension is corrected by adjusting the torque arm length. Technician B states that some rear tandem axle suspensions rotate an eccentric adjustment bolt to correct axle alignment. Who is correct?

TASK B.5

A. A only
B. B only
C. Both A and B
D. Neither A nor B

Answer A is incorrect. Technician B is also correct.

Answer B is incorrect. Technician A is also correct.

Answer C is correct. Both Technicians are correct. Most rear tandem axle suspensions adjust by either varying the length of an adjustable torque arm or by changing the position of the torque arm in relation to the frame bracket by rotating an eccentric bolt or adding shims between the torque arm and the frame mounting bracket.

Answer D is incorrect. Both Technicians are correct.

37. A technician turns the steering wheel to the left until the steering effort increases. Technician A states that if the power steering pump goes into the pressure-relief mode, and the clearance between the axle stop and axle stop adjustment screw is incorrect, the poppet valves require adjustment. Technician B states that if the power steering pump goes into the pressure-relief mode, and the clearance between the axle stop and axle stop adjustment screw is incorrect, the axle stops need to be adjusted. Who is correct?

TASK A.15

A. A only
B. B only
C. Both A and B
D. Neither A nor B

Answer A is correct. Only Technician A is correct. If the power steering pump goes into the pressure-relief mode, a noticeable change in the engine load will occur. The poppet valves require adjustment.

Answer B is incorrect. Adjusting the axle stops will not change the pump operation. The poppet valves require adjustment.

Answer C is incorrect. Only Technician A is correct.

Answer D is incorrect. Technician A is correct.

2013 © Delmar, Cengage Learning

TASK C.3

38. Referring to the figure above, what alignment angle is being checked?

    A. Camber

    B. Caster

    C. KPI (kingpin inclination)

    D. Steering radius

Answer A is incorrect. Camber is the angle formed between a true vertical line and a line drawn through the centerline of the tire.

Answer B is correct. Castor is the forward or rearward tilt of the kingpin at the top. A machinist protractor may be used to check caster.

Answer C is incorrect. Kingpin inclination is the angle formed between a true vertical line and a line drawn through the kingpin axis.

Answer D is incorrect. This is the difference in turning angle in each front wheel during a turn.

TASK B.15

39. A bumping action is felt combined with erratic steering during braking and acceleration. What would be the LEAST LIKELY cause?

    A. A dry fifth wheel

    B. A worn trailer kingpin

    C. A slack in the fifth wheel

    D. A trailer not properly loaded

Answer A is incorrect. A dry fifth wheel can cause erratic steering complaints.

Answer B is incorrect. A worn trailer kingpin (pintle) can cause a bumping action during braking and acceleration.

Answer C is incorrect. Slack in the fifth wheel can cause a bumping action during braking and acceleration.

Answer D is correct. An improperly loaded trailer will not cause bumping action but may, in some cases, cause erratic steering.

40. A driver complains of a front-end shimmy with slight vibrations. Technician A says loose or worn kingpins or kingpin bushings might be the cause. Technician B says that an overloaded trailer might be the cause. Who is correct?

TASK C.1

A. A only

B. B only

C. Both A and B

D. Neither A nor B

Answer A is correct. Only Technician A is correct. A front shimmy with slight vibrations could be caused by a loose kingpin or kingpin bearing because the spindle assembly is not tight on the kingpin. This would allow the spindle and hub assembly to shake while traveling down the road causing a front-end shimmy with a slight vibration.

Answer B is incorrect. Overloading might cause a wander and weave condition because not enough weight is transferred to the steer axle but will not cause front-end shimmy.

Answer C is incorrect. Only Technician A is correct.

Answer D is incorrect. Technician A is correct.

41. A truck cab equipped with an air ride suspension leans to one side. Upon inspection, the technician does not find any air leaks. The most likely cause is:

TASK A.3

A. An inoperative air brake control valve.

B. A blocked cab height control valve intake port.

C. A kinked air hose to the air bag.

D. A defective air compressor.

Answer A is incorrect. An inoperative air control valve would affect both cab air bags.

Answer B is incorrect. A blocked air intake port would not allow air to flow to either bag.

Answer C is correct. A kinked hose to the air bag will not allow air to flow to that air bag. Therefore the component will not function properly causing the lean.

Answer D is incorrect. A defective air compressor would affect the entire air ride system.

42. A technician is replacing the kingpins on a conventional axle configuration. During the kingpin installation, he encounters interference between the kingpin and the bushing. All of the following could be the cause EXCEPT:

TASK B.2

A. The right-side kingpin was switched with the left-side kingpin.

B. A burred bushing during installation.

C. An improperly aligned bushing.

D. Bushings were not reamed.

Answer A is correct. The right- and left-side kingpins are the same.

Answer B is incorrect. A burred bushing will interfere with kingpin installation.

Answer C is incorrect. An improperly aligned bushing will interfere with kingpin installation.

Answer D is incorrect. Omission of the reaming operation will leave burrs that will interfere with the kingpin installation.

TASK B.3

43. When installing shock absorbers, if the torque applied to the mounting hardware is inadequate, all of the following may be a result EXCEPT:

A. Fastener fatigue.
B. Noisy operation.
C. Consistent directional stability.
D. Premature bushing failure.

Answer A is incorrect. Inadequate torquing of the shock absorber mounts can cause fastener fatigue.

Answer B is incorrect. The inadequate torquing of the shock absorber can lead to noisy operation.

Answer C is correct. Inadequate torquing of the shock absorber will not cause consistent directional stability. It could lead to vehicle sway.

Answer D is incorrect. Inadequate torquing of the shock absorber can cause premature bushing failure.

TASK C.1

44. While driving a vehicle at 28 mph, a driver notices a vibration in the steering wheel. Which of the following is the LEAST LIKELY cause?

A. Improper wheel balance
B. Worn kingpin bushings
C. A bent wheel mounting surface
D. A shifted belt inside a tire

Answer A is incorrect. Improper wheel balance would present as a vibration in the steering wheel because the road forces are transmitted through the steering linkage to the steering wheel.

Answer B is correct. Worn kingpin bushings would present steering instability issues, but not a vibration.

Answer C is incorrect. A bent wheel mounting surface would cause the wheel assembly to be out of balance. This would result in road forces to be felt as a vibration in the steering wheel.

Answer D is incorrect. When a belt shifts within a tire, the tire becomes imbalanced. This imbalance condition would present as a vibration at the steering wheel.

TASK A.6

45. A routine inspection shows discoloration of power steering fluid. The LEAST LIKELY cause for this condition might be:

A. The wrong type of fluid.
B. Mixed brands of fluid.
C. Water mixed with fluid.
D. Overheated or burned smell condition.

Answer A is incorrect. Adding the wrong type of fluid would contaminate and discolor the power steering fluid.

Answer B is correct. Power steering fluids from different manufacturers are compatible with each other. Dye is added to identify production runs and fluid identification. This will not discolor existing fluid.

Answer C is incorrect. Water in the power steering fluid will cause the fluid to have a milky color and will also cause foaming of the fluid.

Answer D is incorrect. Overheated fluid will darken the fluid and will appear brownish in color. It will also have a burned smell.

46. While replacing the steering knuckle bushings, a technician finds there is interference during the kingpin installation. The LEAST LIKELY cause would be:

TASK B.2

A. A worn axle eye.
B. A burred bushing during installation.
C. An improperly aligned bushing.
D. Omission of the reaming operation.

Answer A is correct. If the eye of the axle is worn, the kingpin would have more room to float in the eye. This would not cause a binding during installation. However, since there is a slight interference fit between the axle eye bore and the kingpin, the axle eye should be repaired or the axle replaced if a worn condition is found.

Answer B is incorrect. A burred bushing would cause the kingpin to hang up in the bushing bore.

Answer C is incorrect. The bushings should be on the same vertical axis. If one bushing is misaligned, it would cause a binding effect when the kingpin is being installed.

Answer D is incorrect. If the kingpin bushings are not reamed to size after installation, the kingpin would not travel through the bushing.

47. A technician has identified excessive tire wear. Which of the following suspension components is the LEAST LIKELY cause?

TASK D.1

A. A bent spindle
B. An improper tie rod setting
C. A worn center cross tube bushing
D. A preloaded axle wheel bearing

Answer A is incorrect. A bent spindle would cause inside or outside tire tread wear.

Answer B is incorrect. An improper tie rod setting will produce excessive toe-in or toe-out conditions and will be seen as featheredge wear across the tire tread.

Answer C is incorrect. If the cross tube bushings become worn, there would be excessive play in the suspension. This would cause a change in the camber and toe settings on the axle and wear the tires accordingly.

Answer D is correct. Preloading the axle bearing is detrimental to the bearings, but has no affect on tire wear.

48. What is the LEAST LIKELY cause for a driver to complain about rough ride characteristics?

TASK B.10

A. Excessive positive caster
B. A leaking or damaged shock absorber
C. Loose/worn suspension component
D. Defective rear wheel bearings

Answer A is incorrect. Excessive positive caster might cause a rough ride because the load is transmitted too far out in front of the vehicle. The driver would feel every bump in the road.

Answer B is incorrect. If the shock absorbers are leaking or damaged they can't absorb road shock and provide the dampening effect they are designed for.

Answer C is incorrect. Loose or worn suspension would allow for the components to shift and not provide the load cushioning needed for a smooth ride.

Answer D is correct. A defective rear wheel bearing would cause noise, especially while cornering, but have no effect on the ride characteristics of the vehicle.

TASK A.10

49. The LEAST LIKELY cause for a power steering pump pulley to become misaligned is:

A. An over-pressed pulley.

B. A loose fit from the pulley hub to pump shaft.

C. A worn or loose pump mounting bracket.

D. A broken engine mount.

Answer A is incorrect. An over-pressed pulley would change the pulley position on the pump shaft causing a misalignment.

Answer B is incorrect. If the pulley is not a tight fit to the pulley hub, the pulley may move on the hub causing the pulley to become misaligned.

Answer C is incorrect. A worn or loose pump mounting bracket would allow the pump assembly to move out of position causing pulley misalignment.

Answer D is correct. A broken motor mount will not cause this issue because all of the components would still be running in unison with each other. The broken mount might allow for the entire engine assembly to move.

TASK C.3

50. When adjusting caster, what is the LEAST LIKELY procedure to perform?

A. Use only one shim on each side.

B. The width of the shim should equal the width of the spring.

C. Install the tapered shim with the thickest section to the rear of the vehicle.

D. Reverse one shim to correct for a twisted axle.

Answer A is incorrect. Only one shim of proper thickness should be used on each side to prevent shim shifting and breakage.

Answer B is incorrect. The shim width should be the same as the spring width to prevent shifting and ensure proper clamping force is maintained by the U-bolts.

Answer C is incorrect. Since caster is set positive, the thick part of the shim should be toward the rear of the vehicle. Reversing the shim would set the caster angle to a negative setting.

Answer D is correct. Twisted axles should be replaced.

PREPARATION EXAM 4 – ANSWER KEY

1. A
2. B
3. D
4. A
5. D
6. B
7. C
8. B
9. D
10. C
11. D
12. C
13. C
14. C
15. C
16. A
17. D
18. B
19. D
20. B
21. C
22. B
23. D
24. C
25. B
26. C
27. D
28. D
29. A
30. A
31. C
32. C
33. A
34. B
35. D
36. A
37. B
38. D
39. A
40. A
41. C
42. D
43. D
44. C
45. B
46. A
47. D
48. D
49. D
50. B

PREPARATION EXAM 4 – EXPLANATIONS

TASK A.3

1. Technician A says that if the cab air bag has a leak, the pressure protection valve closes to protect the air brake system from air loss. Technician B says the pressure protection valve protects the air bag if the air brake system has excessive pressure. Who is correct?

   A. A only
   B. B only
   C. Both A and B
   D. Neither A nor B

   Answer A is correct. Only Technician A is correct. When a leak occurs in the cab air bag or any pneumatic accessory, the pressure protection valve will close to prevent a loss of available air pressure for sufficient air brake operation.

   Answer B is incorrect. The pressure protection valve will sacrifice air to the cab air bag to protect braking. It is not a pressure relief valve that protects the brake system from excessive pressure.

   Answer C is incorrect. Only Technician A is correct.

   Answer D is incorrect. Technician A is correct.

TASK C.5

2. When performing a rear-axle alignment on a vehicle equipped with a tandem axle, Technician A says that the front axle should be aligned to the frame first. Technician B says that the rear axle should be aligned first. Who is correct?

   A. A only
   B. B only
   C. Both A and B
   D. Neither A nor B

   Answer A is incorrect. The front axle should be aligned to the rear axle, not to the frame.

   Answer B is correct. Only Technician B is correct. The rear axle is aligned to the frame and then the front axle is aligned to it.

   Answer C is incorrect. Only Technician B is correct.

   Answer D is incorrect. Technician B is correct.

TASK B.6

3. Center shaft wear may be caused by all of the following EXCEPT:

   A. Beam end bushing wear.
   B. Beam center bushing wear.
   C. Torque arm bushing wear.
   D. Front spring bushing wear.

   Answer A is incorrect. End bushing wear will cause cross shaft wear.

   Answer B is incorrect. Center bushing wear will cause cross shaft wear.

   Answer C is incorrect. Torque arm bushing wear will cause cross shaft wear.

   Answer D is correct. As with center shaft wear, front spring bushing wear is a result of equalizer beam and torque arm bushing wear. It will not cause center shaft wear.

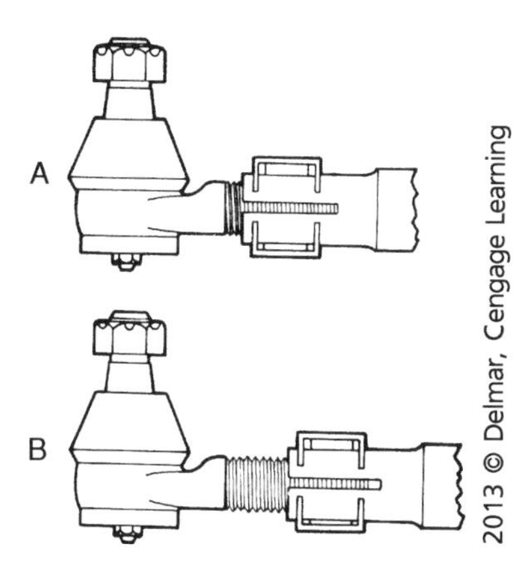

4. Referring to the figure above, you are installing a tie rod end. Which position is the correct installation?

   A. Position A only
   B. Position B only
   C. Both A and B are acceptable
   D. Neither A nor B is acceptable

   Answer A is correct. The tie rod in position A is installed correctly. The threads on the tie rod end should extend past the cross tube split.

   Answer B is incorrect. The threads of the tie rod end do not extend past the split in the cross tube. This may result in damage to the cross tube.

   Answer C is incorrect. Answer A only is correct.

   Answer D is incorrect. Answer A only is correct.

5. After the front springs were replaced on his tractor, the driver complains that the steering wheel does not return to center after a turn but steers fine straight ahead. The tires show no abnormal wear pattern. The first thing to check should be:

   A. Wheel balance.
   B. Camber adjustment.
   C. Toe-out on turns.
   D. Caster adjustment.

   Answer A is incorrect. Wheel balance would cause a bounce or shake.

   Answer B is incorrect. Camber would cause a pull.

   Answer C is incorrect. Toe-out on turns is to prevent tire scuffing.

   Answer D is correct. Caster is a stability angle that helps to return the steering wheel to the straight-ahead position. Caster is the forward or rearward tilt of the kingpin centerline when viewed from the side of the vehicle.

TASK B.11

6. When replacing broken U-bolts, which of the following is the LEAST LIKELY procedure to perform?

A. Check the condition of the leaf spring center bolt.

B. Torque the U-bolt 10 percent above specs to allow for seating and stretch.

C. Clean mounting surfaces and lubricate the U-bolt threads.

D. Re-torque the U-bolts after 1,000 miles of operation.

Answer A is incorrect. Anytime the U-bolts are replaced, the condition of the center bolt should be checked.

Answer B is correct. U-bolts should be torqued to specification when installed. Then the vehicle should be driven and the torque rechecked. Torquing the U-bolts above the manufacturer's specification may damage the U-bolt threads.

Answer C is incorrect. Cleaning the mounting surfaces and lubricating the U-bolt threads ensures that true pull-up torque is obtained when the U-bolts are tightened.

Answer D is incorrect. U-bolt torque should be rechecked after 1,000 miles of operation to ensure that the manufacturer's recommended torque specification is maintained. Spring stack settling may occur and reduce the original U-bolt torque.

TASK A.13

7. A driver complains of reduced power steering assist. Technician A states that the first thing that should be checked is the power steering fluid level. Technician B says the air in the system could be the cause. Who is correct?

A. A only

B. B only

C. Both A and B

D. Neither A nor B

Answer A is incorrect. Technician B is also correct.

Answer B is incorrect. Technician A is also correct.

Answer C is correct. Both Technicians are correct. The first thing to check with power steering system complaints is the fluid level. Air in the system could also affect the steering assist.

Answer D is incorrect. Both Technicians are correct.

TASK D.9

8. A technician is dismounting a split ring tire. When inspecting the components, what is the LEAST LIKELY thing to check?

A. Excessive rust or corrosion buildup

B. Tire brand

C. Bent flanges

D. Deep tool marks on the rings and gutter area

Answer A is incorrect. Excessive rust or corrosion buildup could cause problems when dismounting a split ring tire.

Answer B is correct. Tire brand is insignificant when dismounting a split ring tire as long as the proper tire size and construction are selected to match the manufacturer's rim or wheel rating and size. The diameter of the tire must match the diameter of the rim.

Answer C is incorrect. Bent flanges can be a hazard when dismounting a split ring tire.

Answer D is incorrect. Deep tool marks on the rings and gutter area can be a hazard when dismounting a split ring tire.

9. A technician notices grease leaking out through the pivot bearing while greasing a front-axle kingpin. What should the technician do next?

TASK B.2

   A. Replace the bearing.
   B. Reduce grease injection pressure.
   C. Continue greasing until four drips of grease fall.
   D. Do nothing; it shows thorough distribution of the grease.

   Answer A is incorrect. Grease at the pivot bearing is a normal condition and does not require bearing replacement.

   Answer B is incorrect. Grease at the pivot bearing is a normal condition and does not require a reduction in grease gun pressures.

   Answer C is incorrect. Allowing only four drips of grease is not recommended. The bearing should be greased until fresh grease is present around the entire bearing.

   Answer D is correct. Grease leaking through the pivot bearing is normal when greasing a kingpin.

10. Excessive camber angles may cause all of the following EXCEPT:

TASK C.2

   A. Abnormal tire wear on the inside of the tread.
   B. Abnormal tire wear on the outside of the tread.
   C. Steering instability.
   D. Hard steering going into a turn.

   Answer A is incorrect. Tire wear on the inside of the tread is an indication of excessive negative camber angle.

   Answer B is incorrect. Tire wear on the outside of the tread is an indication of excessive positive camber angle.

   Answer C is correct. Excessive camber will cause the vehicle to pull to the most positive side.

   Answer D is incorrect. Hard steering going into a turn usually indicates excessive positive castor.

11. The drive tires of a vehicle were replaced at time of PM. After only 10,000 miles, they are worn down to the minimum wear indicators. Which of the following could be the cause?

TASK D.1

   A. Damaged/worn rear axle bearings
   B. Rear caster measurement
   C. Front wheel toe-in out of specification
   D. Bent rear axle housing

   Answer A is incorrect. Defective rear axle bearings might cause noise, especially while turning, but will not cause premature rear tire wear.

   Answer B is incorrect. Caster is not a rear wheel measurement.

   Answer C is incorrect. Front wheel toe-in out of specification might cause premature steer axle tire wear, but will not affect premature rear-tire wear.

   Answer D is correct. A bent rear axle housing will cause incorrect camber or toe which would cause premature rear tire wear.

TASK C.4

12. When adjusting the toe setting on a Class 8 tractor, increasing the length of a tie rod has what effect to the toe setting?

    A. Toe-out increases

    B. Toe-in decreases

    C. Toe-in increases

    D. Toe remains neutral

Answer A is incorrect. Toe-out is increased by shortening the length of the tie rod assembly.

Answer B is incorrect. Increasing the length of the tie rod increases toe-in.

Answer C is correct. Increasing the length of the tie rod increases toe-in. Toe setting is generally set for a slight amount of toe-in. This allows for the wheels to obtain a neutral toe setting while in motion.

Answer D is incorrect. Anytime the tie rod assembly is lengthened or shortened, there will be a change to the toe setting. It will not remain neutral.

TASK A.10

13. Technician A says that when inspecting the drive belts on a power steering system, the pulleys should be checked for alignment and wear. Technician B states that when inspecting a power steering system belt, the tension should be relieved and the belt tensioner should be checked for proper operation. Who is correct?

    A. A only

    B. B only

    C. Both A and B

    D. Neither A nor B

Answer A is incorrect. Technician B is also correct.

Answer B is incorrect. Technician A is also correct.

Answer C is correct. Both Technicians are correct. Pulleys should be checked for wear and alignment anytime an inspection of the power steering system is warranted. An operational check of the tensioner should be included in that inspection.

Answer D is incorrect. Both Technicians are correct.

TASK B.8

14. A driver complains of repeated air spring ruptures on a trailer equipped with a lift axle air suspension. Technician A says that the trailer may be continually overloaded. Technician B says that the maximum air pressure adjustment on the hand control valve may be too high. Who is correct?

    A. A only

    B. B only

    C. Both A and B

    D. Neither A nor B

Answer A is incorrect. Technician B is also correct.

Answer B is incorrect. Technician A is also correct.

Answer C is correct. Both Technicians are correct. Both overload and maximum air pressure adjustments may cause air bag ruptures.

Answer D is incorrect. Both Technicians are correct.

15. Technician A says that the objective of a toe specification in truck steering systems is to achieve zero toe when the truck is fully loaded and running at highway speeds. Technician B says that before setting toe the camber angle should be adjusted. Who is correct?

TASK C.4

A. A only
B. B only
C. Both A and B
D. Neither A nor B

Answer A is incorrect. Technician B is also correct.

Answer B is incorrect. Technician A is also correct.

Answer C is correct. Both Technicians are correct. Proper toe adjustments will achieve zero toe when the truck is fully loaded and running at highway speeds. Because camber can change toe settings, it should be checked and adjusted first.

Answer D is incorrect. Both Technicians are correct.

16. Misaligned drive belt pulleys may cause all of the following EXCEPT:

TASK A.10

A. Low pump pressure.
B. Wear on the belt edges.
C. Pulley wear.
D. Belt squeal.

Answer A is correct. Misalignment of the pulleys will not cause low pump pressure.

Answer B is incorrect. Wear on the belt edges may be a result of improper pulley alignment.

Answer C is incorrect. Pulley wear may be a result of improper pulley alignment.

Answer D is incorrect. Belt squeal may be a result of improper pulley alignment.

17. A driver complains of his vehicle leaning in the front end. Upon inspection it is determined that one of the front springs is weak. What is the recommended service procedure?

TASK B.4

A. Align the vehicle using the present ride height.
B. Add an auxiliary spring for added support.
C. Replace the sagged spring only.
D. Replace both springs on the axle as a pair.

Answer A is incorrect. An alignment should not be performed because the ride height would not be correct.

Answer B is incorrect. Auxiliary springs should only be added in pairs and not to correct a spring sag condition.

Answer C is incorrect. If one spring must be replaced on an axle, they should both be replaced.

Answer D is correct. Springs should be replaced in pairs.

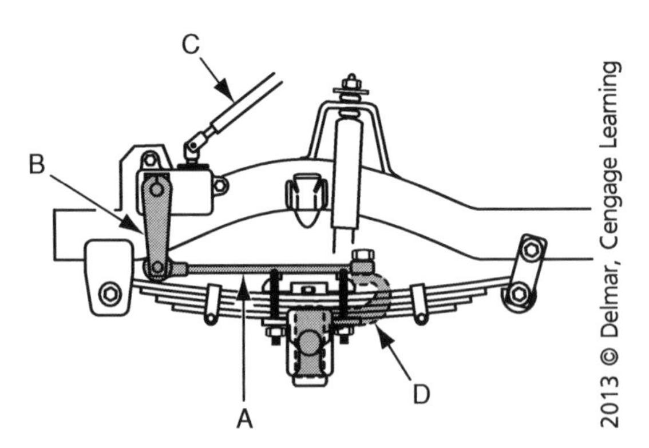

TASK A.16

18. Referring to the figure above, the pitman arm is:

A. Letter A.

B. Letter B.

C. Letter C.

D. Letter D.

Answer A is incorrect. Letter A is the drag link.

Answer B is correct. Letter B is the pitman arm.

Answer C is incorrect. Letter C is the steering shaft.

Answer D is incorrect. Letter D is the steering arm.

TASK C.2

19. What is the LEAST LIKELY effect from an excessive camber angle?

A. Abnormal tire wear on the inside of the tread

B. Abnormal tire wear on the outside of the tread

C. Steering instability

D. Hard steering going into a turn

Answer A is incorrect. Tire wear on the inside of the tread is an indication of excessive negative camber angle.

Answer B is incorrect. Tire wear on the outside of the tread is an indication of excessive positive camber angle.

Answer C is incorrect. Excessive camber at one wheel position on a front steer axle will cause the vehicle to pull to the most positive side.

Answer D is correct. Hard steering going into a turn usually indicates excessive positive castor.

20. A tractor was involved in a front-end collision and the technician suspects that there may be damage to the front axle. To determine if axle damage exists, the technician should lift both front wheels and rotate the tires to:

TASK B.1

A. Measure for radial runout.
B. Measure front wheel setback.
C. Measure relay rod length and compare to axle.
D. Measure toe-in.

Answer A is incorrect. Checking runout will identify a bent rim, not front axle damage.

Answer B is correct. You measure the front wheel setback on both wheels and compare. A difference in measurement could be caused by a bent axle. Setback is a condition in which one wheel is moved rearward in relation to the other front wheel.

Answer C is incorrect. The relay rod length will not determine axle damage.

Answer D is incorrect. Toe-in would only identify a bend in the steering component.

21. The LEAST LIKELY cause of abnormal tire wear, shimmy, or vibration is:

TASK C.1

A. Tire/wheel imbalance.
B. Excessive wheel or hub runout.
C. Excessive cam brake stroke.
D. Improper tire mounting.

Answer A is incorrect. Tire/wheel imbalance may cause a shimmy or vibration.

Answer B is incorrect. Excessive wheel or hub runout may cause a shimmy or vibration.

Answer C is correct. Excessive cam brake stroke may lead to poor brake application and air brake imbalance, but it will not cause abnormal tire wear, shimmy, or vibration.

Answer D is incorrect. Improper tire mounting may cause a shimmy or vibration.

22. Technician A says that lift axles are always a fixed straight axle design. Technician B says that lift axles are designed to be either a fixed axle design or a steerable axle design. Who is correct?

TASK B.8

A. A only
B. B only
C. Both A and B
D. Neither A nor B

Answer A is incorrect. Lift axles are available in both fixed and steerable designs.

Answer B is correct. Only Technician B is correct. Lift axles are available in both fixed and steerable designs.

Answer C is incorrect. Only Technician B is correct.

Answer D is incorrect. Technician B is correct.

TASK A.14

23. When inspecting a power steering gear before removal, which of these tasks would be the LEAST LIKELY task to perform?

 A. Rotate the input shaft and visually determine if it is true.
 B. Clean and inspect for evidence of fluid leakage.
 C. Check all mounting fasteners.
 D. Adjust the truck toe.

Answer A is incorrect. A visual inspection of the input shaft should be performed to determine if the shaft is bent or damaged.

Answer B is incorrect. Cleaning and inspecting for fluid leakage should be performed to determine if any repairs to the pump are necessary prior to installation.

Answer C is incorrect. Checking for proper mounting and fasteners for looseness will be a good indication if there is bracket or thread damage.

Answer D is correct. There is no need to check the toe adjustment since this will not be affected by pump removal or installation.

TASK D.11

24. What is the torque specification for the spindle nut on a PreSet hub?

 A. 50 ft-lbs
 B. 200 ft-lbs
 C. 300 ft-lbs
 D. 750 ft-lbs

Answer A is incorrect. 50 ft-lbs is the initial torque for seating the bearings. The final torque is 300 ft-lbs.

Answer B is incorrect. 200 ft-lbs of torque is not adequate to maintain proper hub retention.

Answer C is correct. The torque specification for the spindle nut on a PreSet hub is 300 ft-lbs.

Answer D is incorrect. 750 ft-lbs is excessive and would most likely damage the hub components and eliminate the hub end-play.

TASK C.3

25. While adjusting front wheel caster:

 A. Excessive positive caster decreases steering effort.
 B. Excessive positive caster may cause front-wheel shimmy.
 C. The front suspension caster becomes more negative when the rear suspension height is lowered.
 D. Excessive negative caster results in harsh ride characteristics.

Answer A is incorrect. Excessive positive caster increases steering effort because the front wheels want to maintain a straight-ahead position. It also increases the downward tilt of the spindle when turning. This causes the suspension and chassis lift to increase, thus increasing the steering effort to overcome gravitational forces.

Answer B is correct. Excessive positive caster will direct the load right at the edge of a pothole or irregularity in the road. This results in road shock being transmitted through the kingpin to the suspension and chassis. This will cause the front wheels to oscillate at lower speeds producing front-wheel shimmy.

Answer C is incorrect. When the rear suspension height is lowered, the axle and spindle assembly will rotate rearward. This increases positive caster.

Answer D is incorrect. Negative caster will improve ride characteristics. Excessive negative caster will decrease directional stability and may cause wander and weave issues.

26. A truck frame is being inspected while on a frame-straightening rack. Technician A says buckle is a condition where one side rail is bent upward from its original position. Technician B says that sag occurs when the frame or one side rail is bent downward from its original position. Who is correct?

TASK B.13

A. A only

B. B only

C. Both A and B

D. Neither A nor B

Answer A is incorrect. Technician B is also correct.

Answer B is incorrect. Technician A is also correct.

Answer C is correct. Both Technicians are correct. Buckle is when one frame side rail is bent upward from its original position. This condition is produced by an impact at the frame end. Frame sag is when one frame side rail is bent downward from its original position. Overloading and/or fatigue usually cause this condition.

Answer D is incorrect. Both Technicians are correct.

27. When airing tires being restrained in a device, all of the following should be used EXCEPT:

TASK D.9

A. A clip-on chuck.

B. An in-line valve with pressure gauge or pre-settable regulator.

C. Enough hose between the clip-on chuck and the in-line valve to allow the user to stand outside of the trajectory.

D. A handheld pressure gauge.

Answer A is incorrect. A clip-on air chuck is required by OSHA to place the technician outside of the tire/wheel trajectory area in case of a sudden tire/wheel separation.

Answer B is incorrect. This is an OSHA requirement. By using an in-line gauge or preset air regulator, the technician may maintain a safe working distance from the tire/wheel assembly in case of a sudden tire separation from the wheel.

Answer C is incorrect. The use of an extended air hose ensures that the technician is out of the tire/wheel trajectory in case of a sudden separation.

Answer D is correct. Never use a hand pressure gauge to check tire pressure on a tire being aired in a restraining device.

TASK C.1

28. The LEAST LIKELY cause of a front steer axle pull condition is:

 A. A dragging brake.
 B. An out-of-adjustment brake.
 C. Incorrect brake timing.
 D. An incorrect crack pressure relay valve.

Answer A is incorrect. The vehicle would pull to the side that the brake is dragging on due to increased rolling resistance on that wheel end.

Answer B is incorrect. If the brakes are not adjusted properly with the same brake chamber push rod stroke, the vehicle will pull to one side when the brake is applied.

Answer C is incorrect. Incorrect brake timing between wheel assemblies on the same axle will cause the vehicle to pull to one side while braking because one brake shoe assembly will apply braking force before the opposite brake assembly does.

Answer D is correct. An incorrect crack pressure at the relay valve will not cause a front steer axle pulling condition.. Crack pressure is the amount of air pressure required to open the relay valve so air will flow through the valve and apply the brakes. The same air pressure would be applied to the brake chambers. The relay valve cannot cause a steer axle pull.

TASK B.3

29. A driver complains of a rough ride and body sway on his tractor. While performing an inspection, a faulty shock absorber is found. Technician A says replace the shock absorbers in pairs. Technician B says if the shock absorbers are not leaking they are still good. Who is correct?

 A. A only
 B. B only
 C. Both A and B
 D. Neither A nor B

Answer A is correct. Only Technician A is correct. Shock absorbers should be replaced in pairs.

Answer B is incorrect. Besides leaking, a shock absorber could be bent, empty, or have a sticking valve.

Answer C is incorrect. Only Technician A is correct.

Answer D is incorrect. Technician A is correct.

30. After performing a front-axle and linkage binding test, the technician notes that the wheel and tire do not return to the straight-ahead position. Which of these components should the technician check?

A. Kingpin bearings
B. Front suspension springs
C. Rear suspension springs
D. Shock absorbers

Answer A is correct. If the kingpin bearings are dry, rough, or seized, they will affect the ability of the spindle to pivot freely around the kingpin. This would inhibit the wheel and tire from returning to the straight-ahead position.

Answer B is incorrect. Front spring suspension issues may affect driveabliity and ride characteristics but would not be a cause for the wheels/tires to not return to the straight-ahead position.

Answer C is incorrect. Rear suspension springs would affect the ride characteristics and may contribute to alignment issues and tire wear.

Answer D is incorrect. The shock absorbers control spring oscillation and would not cause steering issues.

31. While checking the fluid level in a power steering system, Technician A says that foaming in the remote reservoir may indicate air in the system. Technician B says that most original equipment manufacturers (OEMs) recommend that the fluid be warmed up prior to checking the level. Who is correct?

A. A only
B. B only
C. Both A and B
D. Neither A nor B

Answer A is incorrect. Technician B is also correct.

Answer B is incorrect. Technician A is also correct.

Answer C is correct. Both Technicians are correct. Foaming of the fluid indicates air in the system. Also, most OEMs recommend that the fluid be at operation temperature prior to checking the level.

Answer D is incorrect. Both Technicians are correct.

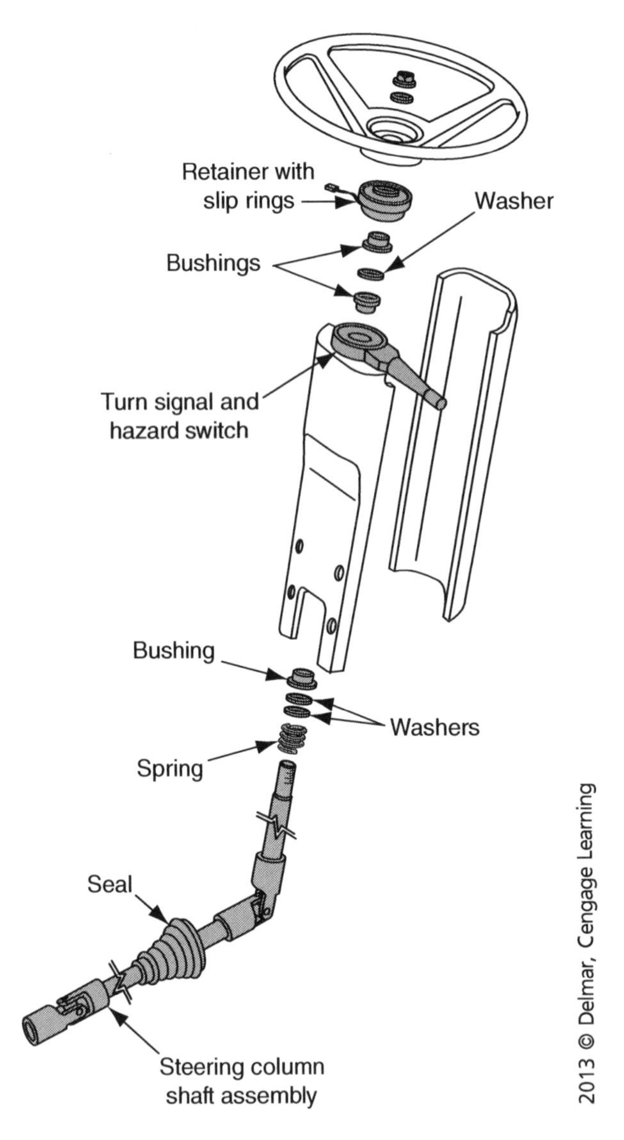

TASK A.1

32. Referring to the figure above, noise in the steering column is being discussed. Technician A says that noise might be caused by the upper or lower shaft bearing being tight or frozen. Technician B says that noise might be caused from a steering shaft snap ring not being seated. Who is correct?

A. A only

B. B only

C. Both A and B

D. Neither A nor B

Answer A is incorrect. Technician B is also correct.

Answer B is incorrect. Technician A is also correct.

Answer C is correct. Both Technicians are correct. Roller and ball bearings are designed to move freely. When they become tight or frozen, they may become noisy due to the increased rolling resistance. The steering shaft snap ring is designed to maintain component positioning within the column. If the steering shaft snap ring is not seated in the groove, components may move creating noise within the column.

Answer D is incorrect. Both Technicians are correct.

33. While turning left or right on a vehicle equipped with power steering, the driver hears a squeaking noise. This might be caused by:

TASK B.2

A. A dry kingpin pivot bearing.
B. A bent front wheel.
C. A fluid reservoir overfilled.
D. Worn front shock absorbers.

Answer A is correct. A dry kingpin pivot bearing can cause a squeak when turning the wheel left to right.

Answer B is incorrect. A bent front wheel will cause a vibration while in motion, but it will never cause a squeaking sound.

Answer C is incorrect. An overfilled fluid reservoir may cause a whining sound but not a squeak.

Answer D is incorrect. Worn shock absorbers may cause a squeaking sound when bounced up and down but not when turning left or right.

34. After mounting a rim on a spoke wheel on a drive axle, a technician checks the lateral runout and finds it exceeding the maximum allowable runout specification of 0.125 inches. The first step to correct this procedure should be:

TASK D.4

A. Remove the wheel, demount the tire, and straighten the wheel in a hydraulic press.
B. Loosen the stud nut at the point of greatest clearance, and tighten the stud nut opposite this nut.
C. Loosen and re-torque all the stud nuts to specifications.
D. Clean, inspect, lubricate, and adjust the wheel bearings.

Answer A is incorrect. You should never attempt to straighten a wheel.

Answer B is correct. The first step in trying to correct runout on a spoke wheel is to loosen the stud nut at the point of greatest clearance and tighten the opposite stud nut. Then be sure all stud nuts are tightened to specifications.

Answer C is incorrect. Re-torquing all of the stud nuts will not correct the problem.

Answer D is incorrect. Wheel bearing inspection and adjustments are done with the tire/wheel assembly removed. In addition, since the maximum wheel bearing end-play specification is 0.005 inches, adjusting the wheel bearing will have little effect on the overall lateral tire/wheel runout.

35. When installing a hub cap, all of the following requirements need to be followed EXCEPT:

TASK D.10

A. Use SAE grade 5 bolts or stronger.
B. Do not use star washers.
C. Do not use split lock washers.
D. Pack the hub cap with grease.

Answer A is incorrect. OEM recommendation require the use of a grade 5 or higher bolt when installing the hub cap.

Answer B is incorrect. The use of any type of locking washer is not recommended. This might allow for oil passage around the washer.

Answer C is incorrect. The use of any type of locking washer is not recommended. This might allow for oil passage around the washer.

Answer D is correct. The hub cap should be installed and then the oil level checked.

TASK C.2

36. Two technicians are discussing front wheel alignment angles. Technician A says the included angle is the sum of the KPI angle and the positive camber angle. Technician B says if the positive camber angle is increased the included angle is decreased. Who is correct?

A. A only
B. B only
C. Both A and B
D. Neither A nor B

Answer A is correct. Only Technician A is correct. The included angle is determined by adding the amount of positive camber angle and the amount of KPI angle together.

Answer B is incorrect. If the amount of positive camber is increased, the included angle also increases because the included angle is the sum of positive camber and KPI angles.

Answer C is incorrect. Only Technician A is correct.

Answer D is incorrect. Technician A is correct.

TASK B.6

37. An axle alignment on a vehicle equipped with a tandem axle with a walking beam suspension is being performed. Technician A says that if there is axial movement at the center cross shaft, the beam bushings are worn and need to be replaced prior to the alignment being performed. Technician B says that the rear axle should be aligned to the frame first, and then the front axle aligned to the rear axle. Who is correct?

A. A only
B. B only
C. Both A and B
D. Neither A nor B

Answer A is incorrect. The center cross shaft is free floating and should have axial (left to right) movement. If there is radial (front to back) movement, an inspection of the beam bushings should be performed prior to alignment.

Answer B is correct. Only Technician B is correct. The rear axle is aligned to the frame and then the front axle is aligned to it.

Answer C is incorrect. Only Technician B is correct.

Answer D is incorrect. Technician B is correct.

TASK A.14

38. Oil leaks may occur in a power steering gear in all of the locations listed EXCEPT:

A. Side cover o-ring.
B. Pitman shaft oil seal.
C. Top cover seal.
D. Reservoir o-ring.

Answer A is incorrect. An oil leak may occur at the side cover o-ring.

Answer B is incorrect. An oil leak may occur at the pitman shaft oil seal.

Answer C is incorrect. An oil leak may occur at the top cover seal.

Answer D is correct. The reservoir is not part of the steering gear.

39. While performing a routine PM inspection, a technician notes that there are cupping marks around the circumference of the tire. The most likely cause of this tire condition is:

TASK D.1

A. Shock absorbers.

B. Camber.

C. Toe-out on turns.

D. Toe-in.

Answer A is correct. Worn shock absorbers can cause cupping marks around the tire. Worn shock absorbers allow the tire to bounce causing a cupping mark in the tire.

Answer B is incorrect. Camber will cause tread to wear unevenly across the tire.

Answer C is incorrect. Toe-out on turns is generally not adjustable and will only cause tread scuffing when a problem occurs.

Answer D is incorrect. Toe-in will generate featheredge tread wear.

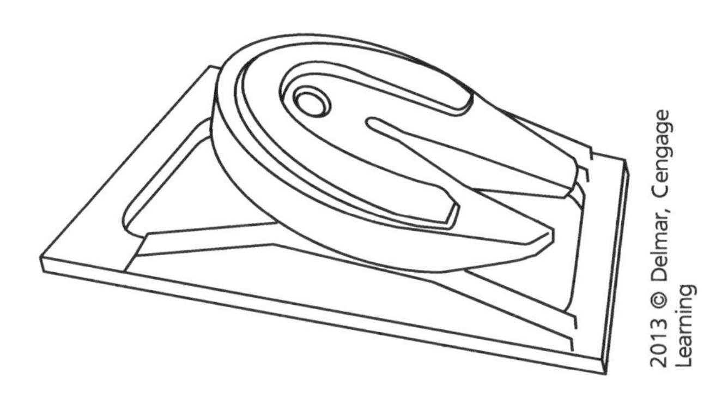

2013 © Delmar, Cengage Learning

40. Referring to the figure above, a technician is lubricating the fifth wheel. What is the recommended type of lube he should use?

TASK B.15

A. Water-resistant lithium-based grease

B. Sodium soap-based grease.

C. Calcium soap-based grease

D. Non-petroleum lubricant

Answer A is correct. Water-resistant lithium-based grease is the recommended lube for the top plate. It has better water-resistance properties compared to sodium soap greases and better high-temperature properties compared to calcium soap greases.

Answer B is incorrect. Sodium-based grease does not have good water-resistance properties and shear capabilities.

Answer C is incorrect. Calcium-based grease breaks down under high temperatures.

Answer D is incorrect. Non-petroleum lubricants are generally not used in heavy truck fifth wheel applications. There are some synthetic greases used in wheel bearing applications.

TASK B.17

41. Technician A says that a bowed upper coupler is caused by trailer overloading. Technician B says that a bowed upper coupler might be the result of using a light gauge steel material. Who is correct?

A. A only
B. B only
C. Both A and B
D. Neither A nor B

Answer A is incorrect. Technician B is also correct.

Answer B is incorrect. Technician A is also correct.

Answer C is correct. Both Technicians are correct. Overloading the trailer will distort the upper coupler. Also, if the gauge of the steel is too thin it cannot support the weight.

Answer D is incorrect. Both Technicians are correct.

TASK C.2, C.3

42. Technician A says that if the rear suspension height is lowered, the front wheel caster becomes more negative. Technician B says that if the camber is negative, the camber angle must be added to the KPI angle to obtain the included angle. Who is correct?

A. A only
B. B only
C. Both A and B
D. Neither A nor B

Answer A is incorrect. Technician A is incorrect.

Answer B is incorrect. Technician B is incorrect.

Answer C is incorrect. Both Technicians are incorrect.

Answer D is correct. Neither Technician is correct. If the rear suspension is lowered, the caster angle becomes more positive, not more negative. If the camber angle is negative, it must be subtracted from the KPI angle, not added, to obtain the included angle.

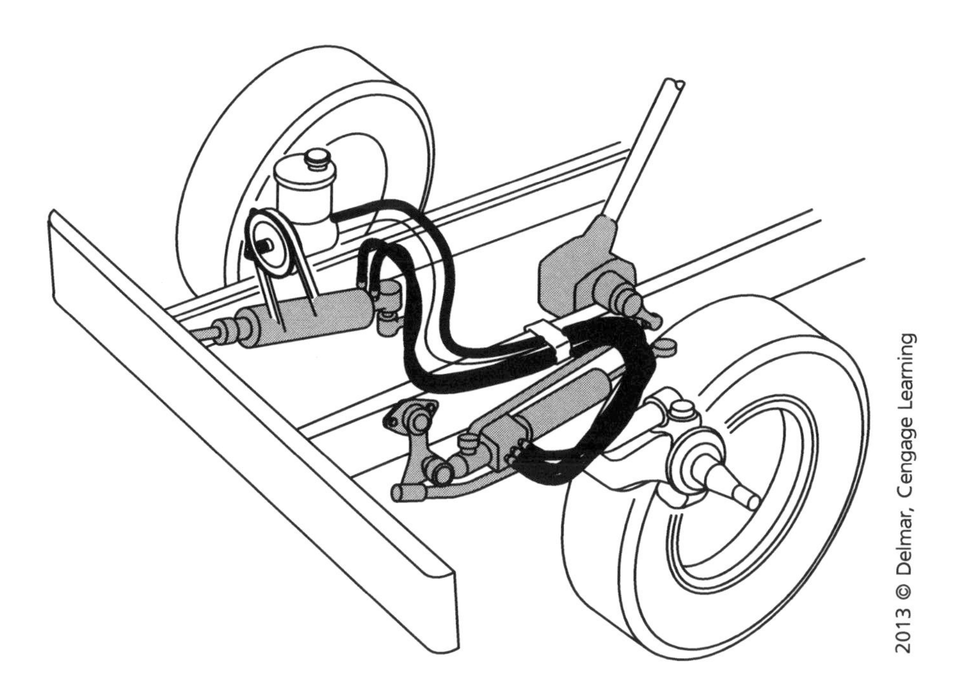

43. The linkage-assist-type power steering shown in the above figure binds when turning corners in either direction, but with short steering corrections, wheel recovery is normal. Which one of these is the LEAST LIKELY cause?

TASK A.13

A. Worn kingpins
B. Improper sector lash adjustment
C. A bent worm gear
D. Worn-out tie rod end assembly

Answer A is incorrect. Worn kingpins can cause binding during cornering because the spindle is not rotating around a true axis anymore and may contact the axle eye.

Answer B is incorrect. Improper steering gear mesh would place excessive force on the steering gear components causing them to bind while cornering.

Answer C is incorrect. A bent worm gear would cause a binding effect within the steering gear.

Answer D is correct. Worn tie rod ends cause the opposite effect and would cause a drifting or hard steering complaint.

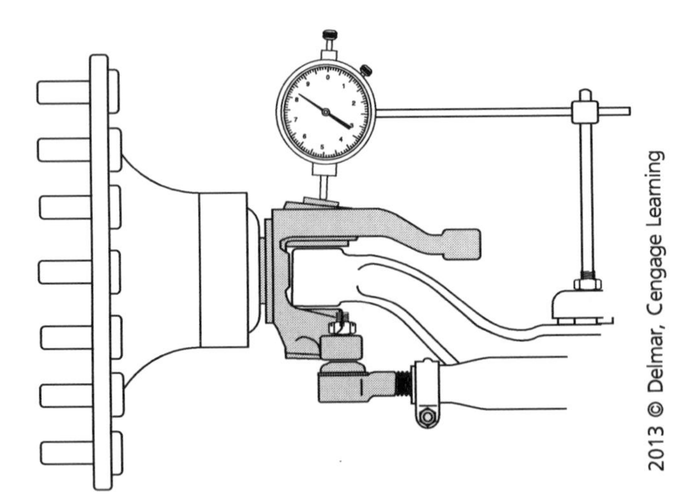

TASK B.2

44. With the dial indicator set up as shown in the above figure, what is being checked?

A. Camber

B. Ball joint end-play

C. Steering knuckle vertical play

D. Lower bearing free-play

Answer A is incorrect. Checking camber requires the use of alignment equipment, not prying downward with a pry bar.

Answer B is incorrect. This front axle uses a kingpin, not ball joints.

Answer C is correct. The technician is checking steering knuckle vertical play. Prying the steering knuckle down and up will produce a play reading on the dial indicator.

Answer D is incorrect. Checking steering knuckle deflection requires moving the knuckle sideways with the dial indicator mounted on the axle to read the radial play.

TASK B.7

45. When the height control lever of the suspension height control valve (leveling valve) is raised upward off of horizontal on an air ride suspension, what happens to the air pressure within the air bags?

A. The air is exhausted.

B. The bag is further inflated.

C. The suspension air circuit is dumped.

D. The chassis system pressure is reduced.

Answer A is incorrect. The air pressure in the air springs is exhausted through the height control valve when the control lever of the suspension height control valve is lowered, not raised, off of horizontal, until the lever resumes a horizontal position.

Answer B is correct. When the height control lever is raised, air pressure is increased in the air bag.

Answer C is incorrect. The leveling valve only controls the air pressure in the air bags and does not affect the entire air suspension circuit.

Answer D is incorrect. The leveling valve only controls the air pressure in the air bags and does not affect the entire chassis system pressure.

46. When discussing end-play on a PreSet hub, Technician A says when the end-play is beyond 0.006 inches the hub requires servicing. Technician B says that a PreSet hub is not serviceable and needs to be replaced. Who is correct?

TASK D.11

A. A only
B. B only
C. Both A and B
D. Neither A nor B

Answer A is correct. Only Technician A is correct. If the end-play exceeds 0.006 inches, the hub should be inspected and then the wheel bearing end-play adjusted to bring the end-play within specification.

Answer B is incorrect. PreSet hubs are fully serviceable and should be inspected and adjusted if the end-play is beyond 0.006 inches.

Answer C is incorrect. Only Technician A is correct.

Answer D is incorrect. Technician A is correct.

47. The LEAST LIKELY cause of hard steering is:

TASK C.1

A. A dry fifth wheel.
B. An overloaded steer axle.
C. A contaminated power steering system.
D. Leaking shock absorbers.

Answer A is incorrect. If the fifth wheel is dry the trailer can't pivot properly. This erratic pivoting action would cause hard steering.

Answer B is incorrect. Overloading of the steering axle exerts excess pressures on the steering components and tire/wheel assemblies. The result is hard steering.

Answer C is incorrect. Contaminants in the power steering system may adversely affect the system operation, which would cause hard steering.

Answer D is correct. Leaking shock absorbers would cause drivability issues such as wheel hop but will not cause hard steering.

48. The steering wheel fails to return to the top tilt position. All of the following might cause the problem EXCEPT:

TASK A.1

A. Bound up pivot pins.
B. A faulty tilt spring.
C. Turn signal wires that are too tight.
D. Worn tilt bumpers.

Answer A is incorrect. Bound up pivot pins will not allow the column to travel, which might cause the steering wheel to fail to return to the top.

Answer B is incorrect. A faulty tilt spring might cause the steering wheel to fail to return to the top.

Answer C is incorrect. If the turn signal wires are too tight, they can restrict the travel, which might cause the steering wheel to fail to return to the top.

Answer D is correct. Worn tilt bumpers will cause noise when tilting the column, but they will not prevent the steering wheel from returning to the top tilt position.

TASK B.5

49. All of the following are methods of adjusting torque arm length while performing an axle alignment EXCEPT:

    A. Shims between the torque arm and front hanger.
    B. Rotating the threaded portion of the torque arm.
    C. Rotating an eccentric bolt to vary the length.
    D. Replace the torque arm with a longer one.

Answer A is incorrect. Placing shims between the torque arm and front hanger is a method for adjusting the torque arm length.

Answer B is incorrect. Rotating the threaded portion of the torque arm is a method for adjusting the torque arm length.

Answer C is incorrect. Rotating an eccentric bolt to vary the length is a method for adjusting the torque arm length.

Answer D is correct. This is not a method for adjusting the torque arm length.

TASK D.4

50. A truck tire is found to have been run in a very underinflated condition. Technician A says that the tire may be inflated without removing it from the truck. Technician B says that before balancing a tire, wheel runout is checked. Who is correct?

    A. A only
    B. B only
    C. Both A and B
    D. Neither A nor B

Answer A is incorrect. When a truck tire has been run in a very underinflated condition, you must remove it from the truck, inspect the casing, and properly remount it.

Answer B is correct. Only Technician B is correct. Before balancing a tire, the technician should check wheel runout because wheel runout will affect the dynamic balance of the wheel assembly. Dynamic balance is the side to side weight distribution across the wheel assembly. If the wheel runout is found to be out of specification, it should be corrected before the balance procedure is performed..

Answer C is incorrect. Only Technician B is correct.

Answer D is incorrect. Technician B is correct.

PREPARATION EXAM 5 – ANSWER KEY

1. A
2. B
3. D
4. A
5. C
6. A
7. B
8. D
9. C
10. B
11. C
12. C
13. D
14. A
15. C
16. D
17. B
18. C
19. B
20. A
21. D
22. B
23. C
24. D
25. D
26. C
27. C
28. D
29. C
30. C
31. A
32. A
33. B
34. B
35. A
36. B
37. A
38. B
39. C
40. D
41. A
42. C
43. D
44. A
45. D
46. C
47. D
48. B
49. C
50. A

PREPARATION EXAM 5 – EXPLANATIONS

TASK D.1

1. All of the following tire wear conditions are considered abnormal EXCEPT:

   A. Tread river erosion.
   B. Excessive toe-in wear.
   C. Excessive toe-out wear.
   D. Excessive shoulder wear.

Answer A is correct. Tread river erosion is considered normal. It is characteristic of the slow wear rate of radial tires on free-rolling axles. It may vary with individual tire tread design and construction. It is common in linehaul operations in which loads are light and turning is infrequent.

Answer B is incorrect. A featheredge wear pattern across the tire tread from the outside of the tire to the inner edge of the tread indicates that the toe angle setting is out of specification. This is not a normal wear pattern and should be addressed.

Answer C is incorrect. Excessive toe-out tire wear exhibits as a featheredge pattern across the tire tread from the inside of the tire to the outer edge of the tread. This is abnormal and indicates that the toe angle setting is out of spec.

Answer D is incorrect. If properly maintained, tires should wear fairly evenly across the tire tread. Excessive shoulder wear might be caused from underinflation, worn suspension components, misalignment, or improper bead seating.

TASK B.4

2. The LEAST LIKELY cause of premature suspension bushing wear is:

   A. A broken leaf spring center bolt.
   B. Excessive rear axle castor.
   C. Excessive wear to adjacent suspension components.
   D. Loose U-bolts.

Answer A is incorrect. A broken center bolt might allow the spring leaves to shift exerting excess pressure on the suspension bushings.

Answer B is correct. Excessive rear-axle castor might cause the driveline operating angle to be incorrect, but it will not cause the suspension to wear prematurely.

Answer C is incorrect. The suspension components are designed to operate as a unit. If adjacent components are worn and not corrected, this might cause premature bushing wear.

Answer D is incorrect. Loose U-bolts would allow the axle to shift placing the suspension out of alignment. This would cause the bushings to wear prematurely.

3. While adjusting front wheel camber:

TASK C.2

   A. If the front wheel has a positive camber angle, the camber line is tilted inward from the true vertical centerline of the wheel and tire.
   B. Excessive positive camber on the front wheel causes premature tire wear on the inside tread of the tire.
   C. Excessive negative camber on a front wheel causes premature wear on the outside of the tire tread.
   D. Improper camber angle on an I-beam front suspension may be caused by a bent axle or spindle.

   Answer A is incorrect. With positive camber, the camber line is tilted outward from the true vertical centerline of the wheel and tire.

   Answer B is incorrect. Excessive positive camber will cause the tire to wear on the outside of the tire tread.

   Answer C is incorrect. Excessive negative camber will cause the tire to wear on the inside of the tire tread.

   Answer D is correct. If the rest of the steering components are checked and no defects noted, improper camber angles may indicate a bent axle or spindle condition.

4. A vehicle with power steering has a high turning effort at idle. At any other RPM, it operates normally. What is the LEAST LIKELY cause of this condition?

TASK A.5

   A. A sticking rotary valve
   B. Excessive wear between the vanes and rotor of the power steering pump assembly
   C. A sticking flow control valve
   D. An obstruction or kink in the fluid return line

   Answer A is correct. The rotary valve directs fluid pressure to the left or right chambers within the steering gear and is mechanically actuated. It would not stick.

   Answer B is incorrect. If excessive wear exists between the pump vanes and rotor, the pump may not produce adequate pressure at idle. Excessive effort to turn would be the result.

   Answer C is incorrect. A sticking flow control valve would affect steering system pressure causing excessive steering effort at idle.

   Answer D is incorrect. If an obstruction to the fluid flow occurs in the return line, more effort would be needed to overcome the resistance due to the slower engine speed.

5. While diagnosing KPI and front spindle movement:

TASK C.2

   A. When the steering wheel is turned, the front spindle movement is parallel to the road surface.
   B. The KPI angle has no effect on steering wheel returning force.
   C. The KPI angle tends to maintain the wheel in a straight-ahead position.
   D. An increase in the KPI angle decreases steering effort.

   Answer A is incorrect. When the steering wheel is turned, each spindle moves through an arc that tries to force the tire into the ground.

   Answer B is incorrect. KPI helps the wheels return to a straight-ahead position because the weight of the vehicle has a tendency to settle to the lowest point of gravity.

   Answer C is correct. One of the functions of KPI is to help maintain the wheels in the straight-ahead position.

   Answer D is incorrect. KPI increases steering effort because the chassis has to lift slightly when turning.

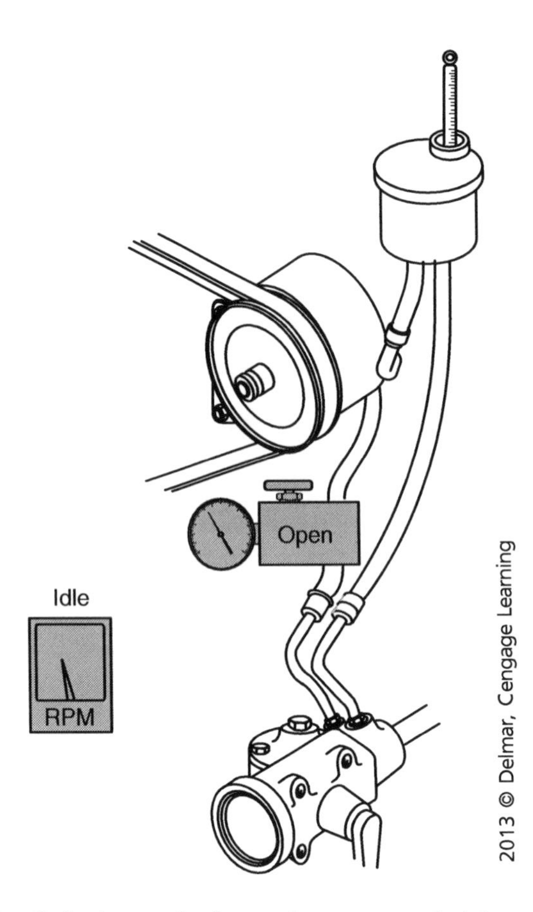

TASK A.8

6. Referring to the figure above, two technicians are discussing the procedure for diagnosing a steering system using a steering system analyzer. Technician A says that when the engine is at idle and the gate valve on the analyzer is open, if the pressure is less than 200 psi there may be a restriction in the high-pressure hose. Technician B says that a reading below 200 psi at idle may indicate that the power steering gear is damaged. Who is correct?

A. A only
B. B only
C. Both A and B
D. Neither A nor B

Answer A is correct. Only Technician A is correct. Low pressure at idle indicates a restriction before the analyzer in the high-pressure line.

Answer B is incorrect. The power steering analyzer is placed between the power steering pump and the power steering gear. Low pressure at the analyzer could not be caused by the steering gear.

Answer C is incorrect. Only Technician A is correct.

Answer D is incorrect. Technician A is correct.

7. An equalizing-beam suspension may also be referred to as:

   A. A torsion bar.
   B. A walking beam.
   C. An air spring.
   D. A multi-leaf variable rate.

   Answer A is incorrect. A torsion bar suspension is a spring suspension consisting of spring steel bars loaded under torsion (twist).

   Answer B is correct. An equalizing-beam suspension may also be referred to as a walking beam suspension.

   Answer C is incorrect. Air spring suspensions are pneumatic suspensions that utilize air springs (bags).

   Answer D is incorrect. Multi-leaf variable rate refers to a type of leaf spring pack used in any number of spring suspension configurations.

8. A driver complains that his tractor bottoms out when hitting a bump. The LEAST LIKELY cause would be:

   A. Excessive weight on the vehicle.
   B. One or more broken leaves in the spring pack.
   C. Weak or fatigued spring assembly.
   D. Broken or missing rebound clips.

   Answer A is incorrect. Excessive weight on the vehicle might exceed the load-carrying capacity of the suspension. If this occurs, the vehicle may bottom out when hitting a bump because the dampening characteristics of the spring assemblies have been neutralized.

   Answer B is incorrect. When one or more leaves in the spring pack are broken, the load-carrying capacity of that spring is reduced. This reduction in load-carrying capacity would result in the vehicle bottoming out when hitting a bump in the road.

   Answer C is incorrect. Weak or fatigued spring assemblies would cause the vehicle to bottom out because the load-carrying capacity and dampening characteristics on the spring have been reduced.

   Answer D is correct. A broken or missing rebound clip would not contribute to the vehicle bottoming out. The rebound clips are there to limit the amount of spring leaf separation that occurs due to spring jounce and rebound characteristics.

9. What is the established industry torque specification for wheel nuts?

   A. 400–450 ft-lbs
   B. 350–400 ft-lbs
   C. 450–500 ft-lbs
   D. 550–600 ft-lbs

   Answer A is incorrect. The established industry torque specification for wheel nuts is 450–500 ft-lbs. Inadequate torque may result in hub damage, wheel damage, and/or broken or damaged wheel mounting hardware.

   Answer B is incorrect. Inadequate torque may result in hub damage, wheel damage, and/or broken or damaged wheel mounting hardware.

   Answer C is correct. The established industry torque specification for wheel nuts is 450–500 ft-lbs. This ensures adequate clamping pressure is applied to the wheel and hub assembly.

   Answer D is incorrect. Over-torquing may result in stud fatigue, wheel damage at the nut contact area of the wheel, and stud thread damage.

TASK C.6

10. While discussing turning radius:

    A. The turning radius is affected by the length of the tie rod.
    B. The turning radius is determined by the steering arm design.
    C. Improper turning radius has no effect on tire tread wear.
    D. During a turn, the inside tire is actually behind the outside tire.

Answer A is incorrect. The length of the tie rod will affect the toe adjustment but not the turning radius.

Answer B is correct. Turning radius is controlled by the steering arms and is specific to chassis wheel base measurements.

Answer C is incorrect. If the turning radius is not correct, the tires will scuff while cornering. This will cause tire tread wear.

Answer D is incorrect. When the vehicle turns, the inside tire is actually ahead of the outside tire because it is traveling on a sharper angle.

TASK A.2

11. Technician A says noise from a manual steering gear might be caused by misalignment of the steering column input shaft. Technician B says that noise coming from the manual steering gear assembly of a linkage-assist-type power steering when turning the steering wheel may be caused by low lubricant level. Who is correct?

    A. A only
    B. B only
    C. Both A and B
    D. Neither A nor B

Answer A is incorrect. Technician B is also correct.

Answer B is incorrect. Technician A is also correct.

Answer C is correct. Both Technicians are correct. A misaligned steering column input shaft can cause binding noises in the steering gear and a steering gear with low lubricant can produce noise during operation.

Answer D is incorrect. Both Technicians are correct.

TASK B.2

12. When using the dial indicator for checking the upper and lower kingpin bushing wear on both conventional and unitized wheel end axles, if the upper bushing is worn or damaged what do you replace?

    A. One bushing in the knuckle
    B. The entire axle
    C. Both bushings in the knuckle
    D. The steering knuckle

Answer A is incorrect. It is never recommended to replace just one bushing in the spindle assembly since the steering stresses are distributed across both bushings. If only one bushing is replaced, it will fail prematurely.

Answer B is incorrect. The spindle houses the bushings.

Answer C is correct. When using the dial indicator for checking the upper and lower kingpin bushing wear on both conventional and unitized wheel end axles, if the upper bushing is worn or damaged, replace both bushings in the knuckle. This maintains steering knuckle integrity and eliminates premature bushing failure.

Answer D is incorrect. The knuckle may be rebushed if no other damage is noted.

13. All of the statements regarding front-wheel toe are true EXCEPT:

A. Driving forces tend to move the front wheels toward a toe-out position on an I-beam front suspension.

B. Improper front-wheel toe causes featheredge tread wear on the front tires.

C. Adjusting the front wheels on an I-beam front suspension to a toe-in position improves directional stability.

D. Front-wheel toe setting on an I-beam front suspension does not affect steering effort.

Answer A is incorrect. Most medium- and heavy-duty vehicles are set with a certain amount of toe-in. This is to compensate for a small amount of lateral movement in the steering linkage. Because the steering linkage is behind the front wheels, the linkage tends to compress when the vehicle is in motion. This will cause the front wheels to toe out.

Answer B is incorrect. Featheredge tire tread wear is the result of improper toe settings. The tires will scuff when in motion and cause this condition.

Answer C is incorrect. One of the purposes for setting the wheels to a slight toe-in setting is so the wheels will be in a straight-ahead position while in motion and will improve directional stability.

Answer D is correct. Not only will tire tread wear be affected by improper toe settings, excessive toe settings will have an adverse effect on directional stability.

14. After measuring kingpin inclination (KPI) on a tractor with an I-beam front axle, the technician finds the KPI on the left side is more than specified. The cause of the problem might be:

A. The knuckle pin may be loose in the end of the axle.

B. The axle may be bent upward in the center of the axle.

C. The steering arm may be loose in the steering knuckle.

D. The drag link may be of wrong length.

Answer A is correct. Loose knuckle pins in the axle eye would allow the spindle to move outward increasing the KPI angle.

Answer B is incorrect. If the axle is bent upward in the middle, the camber angle would change. However, the relationship between the axle eye and the spindle would not be affected.

Answer C is incorrect. The steering arms establish the turning radius and do not have any effect on KPI.

Answer D is incorrect. The drag link connects the pitman arm to the upper steering arm and has no effect on KPI.

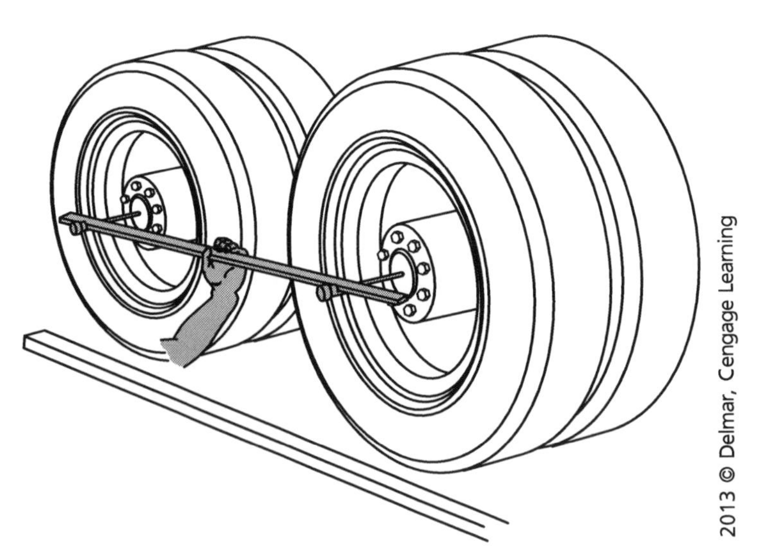

TASK B.5

15. Referring to the figure above, the technician is performing an alignment on a tandem-axle vehicle. He finds the distance between the front and rear tandem axle centers is 0.75 inches (19.05 mm) more on the left side compared to the right side. This could be caused by:

A. A bent rear axle shaft.

B. Worn rear wheel bearings.

C. A bent lower torque rod.

D. An improperly mounted wheel.

Answer A is incorrect. A bent rear axle shaft may cause a vibration and/or noise in the hub or rear carrier assembly, but it will not affect rear axle alignment.

Answer B is incorrect. A loose wheel bearing may cause irregular tire wear, noise, or ABS brake sensor issues, but it would not have that much of an effect on rear axle alignment.

Answer C is correct. A bent lower torque rod could be the cause of these readings. It would change the relative position of the axle housing to the frame rails.

Answer D is incorrect. An improperly mounted wheel may exhibit wheel balance or vibration issues, but it does not have that much of an effect on rear axle alignment.

TASK B.13

16. A frame rail is found to be damaged. Technician A says that in order to reinforce the frame rail, a fishplate frame reinforcement may be used but it should not extend below the frame. Technician B says that a common method for frame reinforcement is to install another section of C-channel, but only on the outside of the frame. Who is correct?

A. A only

B. B only

C. Both A and B

D. Neither A nor B

Answer A is incorrect. Technician A is incorrect. Fishplate frame reinforcements may extend below the frame for added strength.

Answer B is incorrect. Technician B is incorrect. Channel frame reinforcements can be installed on the inside or outside of the frame depending on the strength required and the accessibility of the frame.

Answer C is incorrect. Neither Technician is correct.

Answer D is correct. Neither Technician is correct.

17. The LEAST LIKELY result from negative caster is:

TASK C.3

A. Reduced directional stability.
B. Increased steering wheel returning force.
C. Reduced steering effort.
D. Improved ride quality.

Answer A is incorrect. Reduced directional stability is a result of negative caster. The load is projected behind the kingpin assembly. This will result in a wander and weave complaint.

Answer B is correct. Negative caster reduces the steering wheel's ability to return to center.

Answer C is incorrect. Negative caster reduces the steering effort needed to make a turn because the front wheels will want to move off of center.

Answer D is incorrect. Ride quality is improved with negative caster because the tire and wheel assembly will roll through road irregularities and the road forces are not transmitted as easily to the chassis and suspension.

18. When discussing air lift axles, all of the following are true statements EXCEPT:

TASK B.8

A. The handle tension is not adjustable on the hand control valve.
B. The maximum pressure in the air springs is determined by the axle load and model of the suspension.
C. The minimum pressure setting on the hand control valve determines the air pressure in the air spring with the axle lifted.
D. The maximum pressure setting on the hand control valve determines the air pressure in the air spring with the axle lifted.

Answer A is incorrect. The handle tension is not adjustable on the hand control valve to help maintain air system pressure.

Answer B is incorrect. The original equipment manufacturer (OEM) determines the maximum pressure in the suspension system based on load and suspension design characteristics.

Answer C is correct. The maximum air pressure setting determines the air pressure in the bags.

Answer D is incorrect. The maximum air pressure setting does determine the air pressure in the bags.

19. A technician found the steering arm bent on the driver's side. After replacement, which would be the LEAST LIKELY repair procedure for the technician to perform?

TASK A.18

A. Perform a road test when repairs are completed.
B. Replace both outer tie rod ends.
C. Check and correct changes in wheel alignment.
D. Lube the replacement part after installation.

Answer A is incorrect. The vehicle should be road tested when the job is completed.

Answer B is correct. You only replace defective tie rod ends.

Answer C is incorrect. Whenever a steering component is replaced it is recommended to do a wheel alignment.

Answer D is incorrect. Any replaced part should be lubricated after installation.

TASK D.7

20. A tire and wheel assembly is being removed from the vehicle for inspection. Technician A says if you mark the tire to rim position on removal and reinstall the tire in the same position on the rim, the tire will retain dynamic balance. Technician B says that if a vulcanized repair is made to a tire you do not need to perform a dynamic balance. Who is correct?

A. A only
B. B only
C. Both A and B
D. Neither A nor B

Answer A is correct. Only Technician A is correct. By marking the wheel position, if no repairs are required and the tire position on the rim is not disturbed, then the tire may be returned to its original position and will remain in dynamic balance.

Answer B is incorrect. A vulcanized tire requires rebalancing.

Answer C is incorrect. Only Technician A is correct.

Answer D is incorrect. Technician A is correct.

TASK C.2

21. The LEAST LIKELY condition that would affect camber is:

A. Wheel jounce.
B. Wheel rebound.
C. Road crown.
D. A bent Ackerman arm.

Answer A is incorrect. Wheel jounce is the vertical (upward) wheel movement that occurs when the tire and wheel strike a bump in the road surface. On an I-beam steer axle, if one front tire strikes a bump and wheel jounce occurs, the opposite front wheel temporarily moves to a more positive camber position.

Answer B is incorrect. Wheel rebound is the downward movement of the tire and wheel assembly after wheel jounce has occurred. If a tire and wheel drops into a hole in the road surface on an I-beam steer axle, the opposite front wheel moves to a more negative position.

Answer C is incorrect. Road crown will cause the downhill tire and wheel assembly to become more positive and will cause a pull in the downhill direction.

Answer D is correct. A bent Ackerman arm will not affect camber. It will, however, affect turning radius and tire scrub when cornering.

TASK B.13

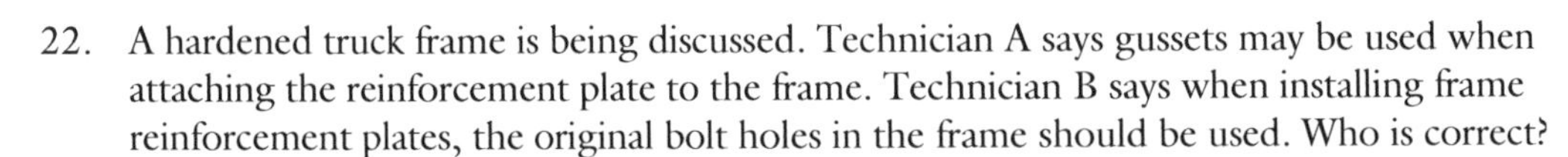

22. A hardened truck frame is being discussed. Technician A says gussets may be used when attaching the reinforcement plate to the frame. Technician B says when installing frame reinforcement plates, the original bolt holes in the frame should be used. Who is correct?

A. A only
B. B only
C. Both A and B
D. Neither A nor B

Answer A is incorrect. You never weld brackets to the frame.

Answer B is correct. Only Technician B is correct. When you install frame reinforcement plates, the original bolt holes in the frame should be used. When the bolt holes are irregularly worn, they may be reamed to fit the next larger size bolt. In addition, they should never be drilled in a vertical line. A staggered bolt pattern with good spacing and sufficient edge distance is preferable.

Answer C is incorrect. Only Technician B is correct.

Answer D is incorrect. Technician B is correct.

23. Excessive worm gear end-play may cause:

TASK A.14

A. Lack of lubrication.

B. A steering fluid leak.

C. Lost motion within the steering gear.

D. Hard steering condition during cold operation.

Answer A is incorrect. Worm gear end-play will not have any effect on the steering gear lubricant.

Answer B is incorrect. Worm gear end-play will not have any effect on a steering fluid leak.

Answer C is correct. Worm gear end-play will produce lost motion within the steering gear. When the input shaft is turned the worm gear must take up the end-play before it can act upon the sector shaft. Proper preloading of the worm shaft bearings is necessary to eliminate worm shaft end-play and prevent steering gear free-play and vehicle wander. A shim adjustment between the lower steering gear cover and the gear housing provides the necessary worm shaft bearing preload.

Answer D is incorrect. Worm gear end-play will not cause cold-weather hard steering. This is usually caused by the use of improper lubricant.

24. Technician A says that when replacing the shaft seal on a power steering pump with an integral reservoir, the pump must be disassembled to facilitate seal installation. Technician B says that when replacing the shaft seal on a power steering pump with a remote reservoir the seal may be serviced without disassembly of the pump. Who is correct?

TASK A.9

A. A only

B. B only

C. Both A and B

D. Neither A nor B

Answer A is incorrect. A seal on an integral reservoir-style pump may be accomplished without disassembly of the pump.

Answer B is incorrect. If the pump has an external reservoir, the pump must be disassembled for seal replacement.

Answer C is incorrect. Neither Technician is correct.

Answer D is correct. Neither Technician is correct. On pumps with an integral reservoir, the pump shaft seal is in front of the shaft bearing and is retained by a snap ring. The seal may be replaced by removing the snap ring and prying the old seal out of the bearing bore. On pumps with an external reservoir, the pump shaft seal is located behind the shaft bearing. In this case, the pump must be disassembled to gain access to the seal.

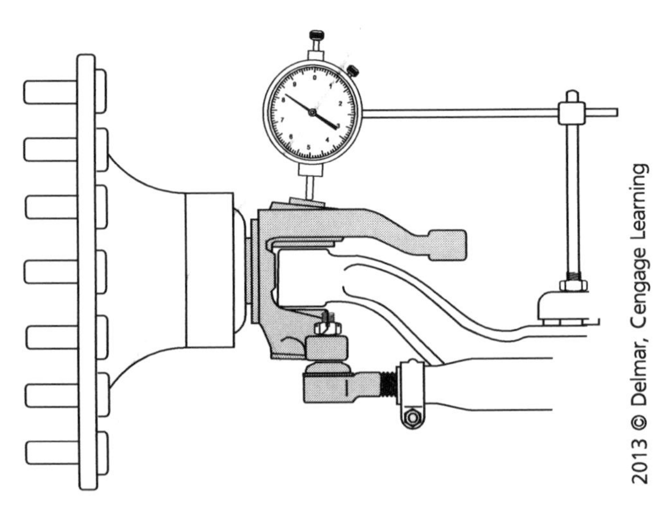

TASK B.2

25. Referring to the figure above, when performing steering knuckle vertical end-play dial indicator checks for in-service axles, the reading must be between which of the following?

A. 0.100 and 0.065 inches

B. 0.010 and 0.065 inches

C. 0.001 and 0.650 inches

D. 0.001 and 0.065 inches

Answer A is incorrect. A reading as high as 0.1 is out of the established specification and would indicate a problem with the spindle and kingpin assembly. Further inspection would be indicated.

Answer B is incorrect. The minimum clearance is 0.001 not 0.010.

Answer C is incorrect. 0.650 is almost 5/8-inch play. The thrust bearing would have to be missing to allow this much end-play.

Answer D is correct. The correct end-play reading is between 0.001 and 0.065 inches.

TASK D.9

26. The rear wheel and hub have been removed as an assembly. The LEAST LIKELY task to be performed when reinstalling the rear wheels and hubs is:

A. Use a wheel dolly to install the dual wheel assembly.

B. Use the OEM-recommended lube on the spindle.

C. Pack the wheel bearings and hub cavity with grease.

D. Set the wheel bearing end-play after installation.

Answer A is incorrect. You do use a wheel dolly to install the dual wheel assembly.

Answer B is incorrect. Certain lubricants are not compatible with other lubricants. Always use the lubricant specified by the OEM.

Answer C is correct. Rear wheel bearings are not packed with grease. They are lubricated from the rear carrier assembly. After the hub is installed, the opposite side of the rear carrier should be raised to allow for lubricant to flow to the installed hub. Once this is complete, the rear carrier should be lowered and the oil level in the carrier checked.

Answer D is incorrect. Because the hub was removed with the wheel assembly, wheel bearings will need to be adjusted to ensure proper end-play at the hub assembly is maintained.

27. A driver complains of a rattling noise on the right side of a medium-duty truck when driving over bumps. Which of these is the most likely cause?

TASK B.10

A. An overtorqued shock absorber mount
B. An underinflated right-rear tire
C. A broken leaf spring rebound clip
D. A cracked air brake supply line

Answer A is incorrect. Only a loose shock absorber mount would rattle over bumps.

Answer B is incorrect. Underinflated tires will cause tire wear, but will not cause a rattling sound.

Answer C is correct. A broken leaf spring rebound clip will rattle when going over bumps due to spring leaf oscillation and not being secured to the spring assembly.

Answer D is incorrect. A cracked air supply line will hiss, not rattle.

28. When checking the tie rod boot for cracks, tears, or other damage, also check the retaining nut to ensure which component listed below is installed.

TASK A.17

A. Boot protector
B. Snap ring
C. Clevis pin
D. Cotter pin

Answer A is incorrect. The boot protector covers the tie rod stud assembly.

Answer B is incorrect. There is no snap ring in this assembly.

Answer C is incorrect. Clevis pins are not used in this assembly.

Answer D is correct. The retaining nut is castellated and requires a cotter pin.

29. After rear suspension work was performed, a truck is now experiencing driveline vibrations. Technician A states that loose U-bolts may be causing the vibration. Technician B says that incorrect installation of the axle shims could cause the vibration. Who is correct?

TASK B.12

A. A only
B. B only
C. Both A and B
D. Neither A nor B

Answer A is incorrect. Technician B is also correct.

Answer B is incorrect. Technician A is also correct.

Answer C is correct. Both Technicians are correct. Loose U-bolts can allow movement of the rear axle housing, which will affect driveline angles. Axle shims that are left out or installed backward will also affect the driveline operating angles causing a vibration.

Answer D is incorrect. Both Technicians are correct.

TASK C.2

30. While discussing wheel alignment variables, Technician A says that road crown is a variable that affects wheel alignment. Technician B says that vehicle loads and unequal weight distribution will affect wheel alignment. Who is correct?

A. A only
B. B only
C. Both A and B
D. Neither A nor B

Answer A is incorrect. Technician B is also correct.

Answer B is incorrect. Technician A is also correct.

Answer C is correct. Both Technicians are correct. Road crown will affect the camber angle. Road crown increase will cause the camber angle to become more positive on the right wheel in relationship to a true horizontal line. The vehicle will then have a tendency to drift or pull to the right. Unequal weight distribution will cause the chassis to lean more to one side or the other and will change the camber setting.

Answer D is incorrect. Both Technicians are correct.

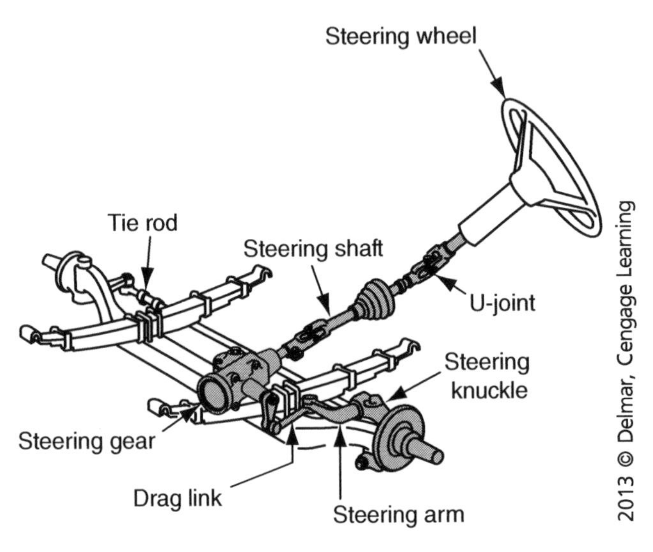

TASK A.2

31. Referring to the figure above, two technicians are discussing excessive steering wheel free-play. Technician A says a worn steering shaft universal joint could cause this free-play. Technician B says that the cause could be a bent idler arm. Who is correct?

A. A only
B. B only
C. Both A and B
D. Neither A nor B

Answer A is correct. Only Technician A is correct. A worn steering shaft universal joint will cause excessive steering wheel free-play by requiring extra steering wheel movement before the steering linkage is affected.

Answer B is incorrect. A bent idler arm will not produce looseness.

Answer C is incorrect. Only Technician A is correct.

Answer D is incorrect. Technician A is correct.

32. During the balancing of a truck tire, a heavy spot is found on the outside edge of the tire. Technician A says when using a static balancing process, a wheel weight is installed 180 degrees from this heavy spot on the outside edge. Technician B says the technician does not have to enter any information in an electronic wheel balancer. Who is right?

TASK D.6

A. A only
B. B only
C. Both A and B
D. Neither A nor B

Answer A is correct. Only Technician A is correct. When using a static balancing process, a wheel weight is installed 180 degrees from this heavy spot on the outside edge. By installing the wheel weight 180 degrees from the heavy spot, the weight counters the heavy spot in the tire.

Answer B is incorrect. The technician must enter the wheel diameter, width, and offset in an electronic wheel balancer.

Answer C is incorrect. Only Technician A is correct.

Answer D is incorrect. Technician A is correct.

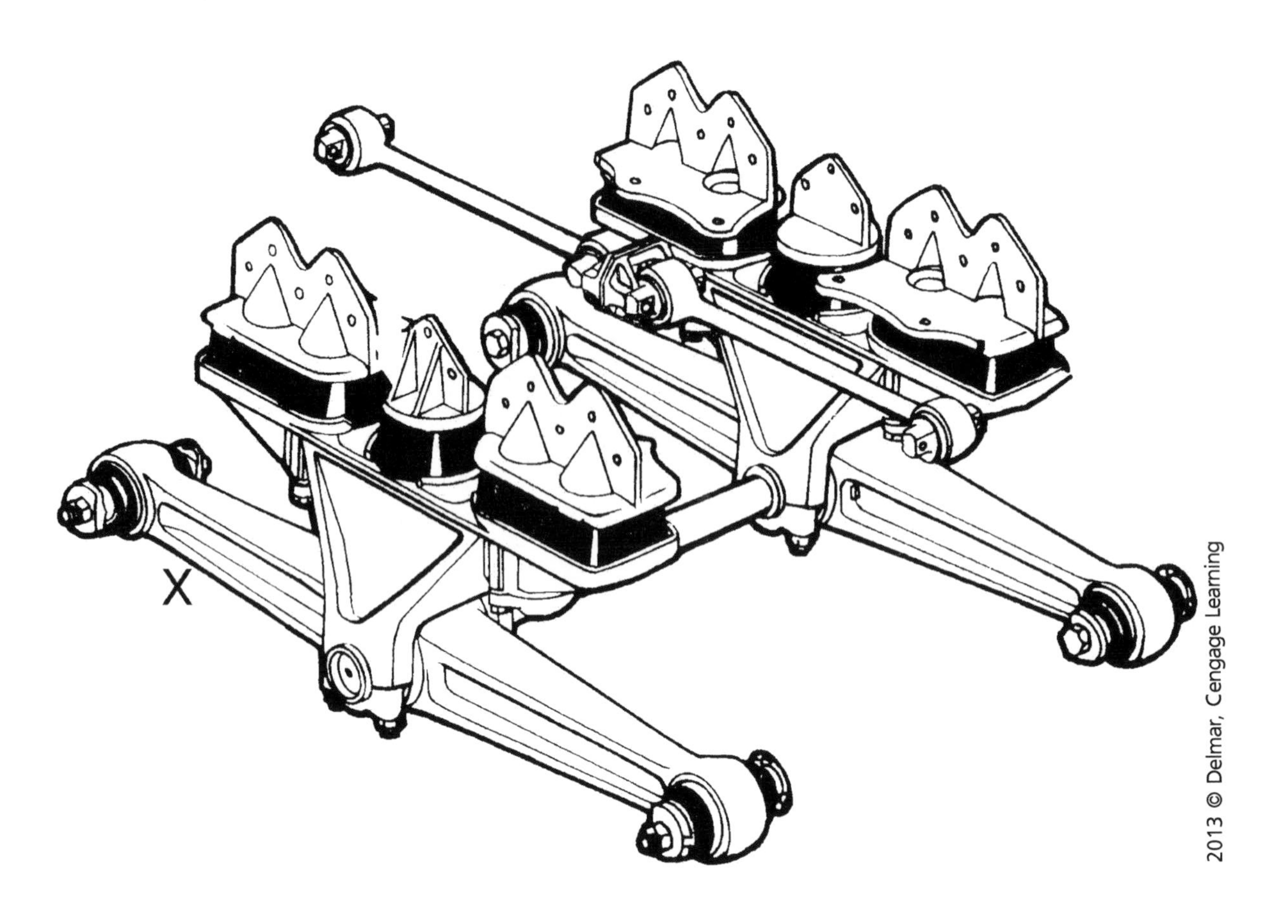

TASK B.6

33. The figure above shows a rubber load cushion-type equalizing beam suspension. Which of the following best describes how driving, braking, and cornering forces are transmitted to the frame?

A. Transmitted directly through the rubber cushions to the frame brackets

B. Transmitted through rubber-bushed vertical drive pins in each cushion

C. Torque rods transmit 100 percent of these forces

D. Equalizing beams transmit 100 percent of these forces

Answer A is incorrect. The rubber cushions act the same way as leaf springs in supporting the load and absorbing road shock.

Answer B is correct. All driving, braking, and cornering forces are transmitted to the frame through rubber-bushed vertical drive pins in each cushion.

Answer C is incorrect. Torque rods control axial forces.

Answer D is incorrect. The beams allow a pair of axles to balance the forces they are subjected to and maintain good tire to road contact.

34. Two technicians are discussing toe-out on turns. All of the statements are true EXCEPT:

TASK C.6

A. Toe-out on turns is the turning angle of the wheel on the inside of the turn compared with the turning angle of the wheel on the outside of the turn.

B. When the front wheel on the inside of the turn has turned 18 degrees outward, the front wheel on the outside of the turn may have turned 20 degrees.

C. Turning radius is the amount of toe-out on turns.

D. Toe-out on turns is determined by the steering arm design.

Answer A is incorrect. This is the definition of toe-out on turns. When cornering, the front steer tires must travel on their own arc. This action is necessary because the inside wheel is actually ahead of the outside wheel.

Answer B is correct. When making a turn, the inside wheel turns at a sharper angle than the outside wheel. Therefore, if the front wheel on the inside of the turn has turned 18 degrees, the outer tire turning angle has to be less than 18 degrees not greater than 18 degrees.

Answer C is incorrect. The turning radius determines the degree of angle each front wheel will travel when cornering. The sharper inside wheel angle during a turn causes the inside wheel to toe-out.

Answer D is incorrect. The steering arms are designed to set the turning radius of the vehicle.

35. What is the proper method for checking wheel end-play?

TASK D.10

A. Remove wheels and drum and measure using a dial indicator.

B. Keep wheels and drum on the vehicle and measure using a dial indicator.

C. Remove wheels and drum and measure using a tape measure.

D. Remove wheels and drum and measure by feel.

Answer A is correct. Removing the wheel assembly allows for a more accurate reading with a dial indicator.

Answer B is incorrect. If the wheel and drum are left on the hub, an inaccurate reading on the dial indicator may occur due to more leverage being applied.

Answer C is incorrect. Wheel end-play is measured in thousandths of an inch. A tape measure cannot be used to measure to that standard.

Answer D is incorrect. Removing the wheel assembly allows for a more accurate reading. However, feel is not a means of measuring free-play. The hub must be checked with a dial indicator.

36. A driver complains that the steering wanders while driving straight down the road. This may be caused by:

TASK A.5

A. A tight worm shaft bearing preload.

B. A loose pitman arm shaft over center preload adjustment.

C. A binding universal joint connected to the steering gear.

D. Contaminated power steering fluid.

Answer A is incorrect. A tight preload would cause excessive steering effort.

Answer B is correct. Inadequate preload allows the worm gear to wander.

Answer C is incorrect. U-joints are static while going straight down the road.

Answer D is incorrect. Contamination of the fluid will not affect steering control while traveling straight ahead. It will affect pump and gear component wear.

TASK C.2

37. Kingpin inclination (KPI) is being discussed. Technician A says that KPI is the inward tilt of the axle eye in relationship to the true vertical tire centerline as viewed from the front. Technician B says that if the KPI measurement is out of specification on an I-beam front axle, it may be adjusted. Who is correct?

    A. A only
    B. B only
    C. Both A and B
    D. Neither A nor B

    Answer A is correct. Only Technician A is correct. KPI is built into the axle and is the amount of inward tilt of the axle eye from a true vertical line. It is used to decrease the amount of camber needed at the wheel end.

    Answer B is incorrect. KPI is not adjustable. It is machined into the axle. If KPI is out of specification, it is an indication of a bent axle. Most axle manufacturers recommend that the axle be replaced if it is bent.

    Answer C is incorrect. Only Technician A is correct.

    Answer D is incorrect. Technician A is correct.

TASK B.2

38. After the first 6,000 miles of new vehicle operation, the draw key nuts should be tightened to which of the following?

    A. 15–30 lb-ft
    B. 30–45 lb-ft
    C. 45–60 lb-ft
    D. 60–70 lb-ft

    Answer A is incorrect. This level of torque is insufficient, which may allow for draw key movement.

    Answer B is correct. The correct torque is 30–45 lb-ft. This torque ensures that the draw key is properly positioned and the kingpin is retained in the axle eye correctly.

    Answer C is incorrect. This level of torque is excessive and may distort the draw key and the dray key bore in the axle.

    Answer D is incorrect. Over-torque may lead to dray key damage or breakage due to excessive stress placed on the draw key threads.

39. Two technicians are discussing split rim tire and wheel assemblies. Technician A says that dismounting and mounting split side rims is extremely dangerous and should only be performed by trained professionals according to OSHA rules and regulations. Technician B says to never hammer on split-side rim rings. Who is correct?

TASK D.3

A. A only
B. B only
C. Both A and B
D. Neither A nor B

Answer A is incorrect. Technician B is also correct.

Answer B is incorrect. Technician A is also correct.

Answer C is correct. Both Technicians are correct. OSHA has established rules and procedures for dismounting and mounting of split rim tire and wheel assemblies. If these procedures are not followed, rim separation may occur and serious injury or death may result. You never use a hammer on split side rim rings. Damage to the ring or ring seat area of the rim may occur, which would result in improper ring seating. Rim separation may occur resulting in serious injury or death.

Answer D is incorrect. Both Technicians are correct.

40. When discussing disk wheels, all of the following statements are true EXCEPT:

TASK D.3,
TASK D.9

A. Disk wheels may have a brake drum that may be removed before the hub and bearings.
B. Disk wheels are mounted on the same bolts as the brake drums.
C. Disk wheels may be retained with wheel nuts having a ball seat.
D. Disk wheels may have retaining bolts that act as pilots to position the wheel.

Answer A is incorrect. Many hub assemblies in use today have outboard brake drums. This allows for the drum to be removed separately from the hub.

Answer B is incorrect. Both the brake drum and the disk wheel are mounted on the same wheel studs. The studs are secured in the hub assembly.

Answer C is incorrect. BUDD-style wheels incorporate wheel nuts that have a ball seat machined in the nut. The ball seat engages a chamfered seat machined into the wheel and centers the wheel on the hub.

Answer D is correct. The retaining bolts have no effect on centering the wheel to the hub. The wheels are centered by the ball seat on the nut in the case of BUDD wheels and by pilot pads machined into the hub on a MOTOR wheel assembly.

TASK C.2,
TASK C.3

41. The steering on a tractor pulls to the left while driving straight ahead and there is no indication of excessive tire wear. Technician A says that the left-front wheel may have more positive caster compared to the right-front wheel. Technician B says the left-front wheel may have excessive positive camber. Who is correct?

A. A only
B. B only
C. Both A and B
D. Neither A nor B

Answer A is correct. Only Technician A is correct. A vehicle will pull to the side with the most positive caster. Caster is also not a tire wearing angle.

Answer B is incorrect. A vehicle with excessive positive camber would cause the pull condition described. However, excessive positive camber would also cause tire tread wear on the outside edge of the tread. Since no tread wear was noted. This would eliminate excessive positive camber as a cause.

Answer C is incorrect. Only Technician A is correct.

Answer D is incorrect. Technician A is correct

TASK B.13

42. All of these statements about truck frame defects are true EXCEPT:

A. A frame sag occurs when the frame rails are bent downward in relation to the rail ends.
B. A frame bow occurs when one or both frame rails are bent upward in relation to the ends of the rails.
C. Vehicle tracking is not affected when one frame rail is pushed rearward in relation to the opposite frame rail.
D. Frame twist occurs when the end of one frame rail is bent upward or downward in relation to the opposite frame rail.

Answer A is incorrect. Frame sag occurs when the frame rails are weakened due to overloading. The middle of the frame rails bow downward in the middle.

Answer B is incorrect. When the frame rail buckles, the frame rails are pushed upward. This may result from operating a dump truck with the box up and loaded, snowplow operations, and/or unequal loading of the frame.

Answer C is correct. Diamond frame conditions will cause the axles to run at an angle to the frame. This will cause the vehicle to “dogtrack” while the vehicle is being driven in a straight line.

Answer D is incorrect. Twist damage occurs when frame rails are twisted off a level plane in relation to each other. Accidents, aggressive docking, and snow plow operation may contribute to frame twist.

43. A driver complains of noise coming from the steering gear. The LEAST LIKELY cause for steering gear noise is:

TASK A.5

A. A loose pitman shaft cover preload adjustment.
B. Cut or worn rings on the spool valve.
C. Loose steering gear mounting bolts.
D. Tight steering shaft U-joints.

Answer A is incorrect. A loose pitman shaft cover preload adjustment may cause a rattling noise when the steering wheel is turned.

Answer B is incorrect. Cut or worn rings on the spool valve will produce a squawking noise during a turn.

Answer C is incorrect. Loose steering gear mounting bolts will produce a noise when the wheels are turned.

Answer D is correct. Tight steering shaft U-joints will not cause noise in the steering gear. They may cause binding of the steering shaft and increased steering effort.

44. Repairs are being performed on a medium-duty truck with a rear air suspension. Technician A says that a trammel bar can be used to measure from a straightedge placed at 90 degrees to the frame rail to the front drive axle. Technician B says if the distance from one side of the axle to a fixed straightedge is different, replace the air spring on that side. Who is correct?

TASK C.5

A. A only
B. B only
C. Both A and B
D. Neither A nor B

Answer A is correct. Only Technician A is correct. When checking rear axle alignment, a straightedge may be placed across the frame rails forward of the front drive axle and at 90 degrees to the frame rail. A trammel bar is used to measure the distance from the straightedge to the center of one front drive axle hub. The trammel bar is then used to check the measurement from the straightedge to the center of the other drive axle hub. The measurement from side to side should be equal.

Answer B is incorrect. If the measurement from side to side is different, the axle position may be adjusted through the torque arm or torque leafs.

Answer C is incorrect. Only Technician A is correct.

Answer D is incorrect. Technician A is correct.

45. Two technicians are discussing turning radius. All of these statements are true EXCEPT:

TASK A.20, TASK C.6

A. Turning radius is controlled by the Ackerman arms.
B. You adjust stop bolts to limit it.
C. It will affect tire tread wear.
D. The inside wheel is parallel to the outside wheel.

Answer A is incorrect. The turning radius is designed into the Ackerman arms, which determines the turning radius for the vehicle wheel base.

Answer B is incorrect. The adjustable wheel stops determine the turning radius limits. Turning angle or radius is the degree of movement from a straight-ahead position of the front wheels to either an extreme right or left position.

Answer C is incorrect. Improper turning radius will cause the tires to scrub while cornering, which will affect tire wear.

Answer D is correct. The inside wheel turns at a different angle than the outside wheel because it travels a shorter distance than the outside wheel.

TASK B.16

46. Technician A says that when a sliding fifth wheel is moved forward on a tractor, the percentage of total vehicle weight supported by the trailer is reduced. Technician B says that when a tractor sliding fifth wheel is moved rearward, weight is removed from the steer axle. Who is correct?

A. A only
B. B only
C. Both A and B
D. Neither A nor B

Answer A is incorrect. Technician A is also correct.

Answer B is incorrect. Technician B is also correct.

Answer C is correct. Both Technicians are correct. When the fifth wheel is moved forward, a percentage of the total vehicle weight supported by the trailer is reduced. This is because the weight is transferred to the tractor suspension and axles. In addition, if the fifth wheel is moved rearward, weight is removed from the steer axle.

Answer D is incorrect. Both Technicians are correct.

TASK D.6,
TASK D.7

47. Two technicians are discussing wheel balancing. All of the statements are true EXCEPT:

A. When a wheel and tire have proper static balance, the weight is distributed equally around the axis of wheel and tire rotation.
B. Improper static wheel balance causes wheel tramp.
C. When a tire and wheel have proper dynamic wheel balance, the weight of the tire and wheel is distributed equally on both sides of the wheel center.
D. Improper static wheel balance causes wheel shimmy.

Answer A is incorrect. When a tire is statically balanced, the weight of the tire and wheel is distributed equally around the center of the tire. This allows the tire to rotate without being influenced by centrifugal forces.

Answer B is incorrect. The heavy spot of a statically imbalanced tire and wheel assembly are influenced by centrifugal forces when they are in motion. This influence attempts to move the heavy spot on a tangent line away from the wheel axis lifting the wheel assembly off of the road surface. As the heavy spot moves downward, the tire strikes the road surface with a pounding action. This action of raising and lowering is called wheel tramp.

Answer C is incorrect. Dynamic wheel balance distributes the weight of the tire and wheel from side to side across the tire tread. A dynamically imbalanced tire and wheel will cause a wheel shimmy.

Answer D is correct. Wheel shimmy is caused by a dynamic imbalance. Static imbalance produces wheel tramp.

48. A front-end alignment is being performed on a heavy-duty tractor with an I-beam front axle. Technician A checks and corrects the toe setting first before checking and adjusting any other alignment angle. Technician B says that when using a tram bar and tape, you should make the measurement between the tires at both the front and rear of the tire and the measurement should be taken at spindle height. Who is correct?

TASK C.4

A. A only
B. B only
C. Both A and B
D. Neither A nor B

Answer A is incorrect. The toe setting should be the last alignment check performed. If camber needs to be adjusted, this might affect the toe setting.

Answer B is correct. Only Technician B is correct. Toe settings should be checked at the front and rear of the tire and at the height of the spindle to achieve a static reading.

Answer C is incorrect. Only Technician B is correct.

Answer D is incorrect. Technician B is correct.

49. While discussing collapsible steering column damage, Technician A says that if the vehicle was involved in a front-end impact, the injected plastic inside of the gearshift tube may be damaged. Technician B says that if the injected plastic is sheared inside the steering shaft, the shaft must be replaced. Who is correct?

TASK A.1

A. A only
B. B only
C. Both A and B
D. Neither A nor B

Answer A is incorrect. Technician B is also correct.

Answer B is incorrect. Technician A is also correct.

Answer C is correct. Both Technicians are correct. Front impacts may transmit forces through the steering components and damage the gear shaft. If the injected plastic is sheared, the shaft must be replaced.

Answer D is incorrect. Both Technicians are correct.

50. Technician A says that when installing external components to the frame, a technician should use existing bolt holes whenever possible. Technician B says that if new holes must be drilled, avoid drilling holes close to the neutral fiber of the frame rail. Who is correct?

TASK B.14

A. A only
B. B only
C. Both A and B
D. Neither A nor B

Answer A is correct. Only Technician A is correct. When installing external components to the frame, a technician should use existing bolt holes whenever possible.

Answer B is incorrect. When drilling, you should drill holes as close to the neutral fibers as possible.

Answer C is incorrect. Only Technician A is correct.

Answer D is incorrect. Technician A is correct.

PREPARATION EXAM 6 – ANSWER KEY

1. B
2. A
3. A
4. D
5. D
6. C
7. C
8. C
9. C
10. A
11. D
12. C
13. C
14. A
15. C
16. A
17. A
18. A
19. C
20. B
21. C
22. A
23. D
24. C
25. A
26. B
27. A
28. A
29. C
30. D
31. C
32. D
33. B
34. C
35. C
36. C
37. A
38. A
39. A
40. B
41. D
42. B
43. B
44. C
45. A
46. A
47. B
48. D
49. B
50. A

PREPARATION EXAM 6 – EXPLANATIONS

TASK D.6,
TASK D.7

1. What is the LEAST LIKELY condition that improper wheel balance would cause?
   A. Cupped tire tread wear
   B. Vibration while braking
   C. Wheel hop or tramp
   D. Wheel shimmy

   Answer A is incorrect. Cupped tire tread wear is a result of a static imbalance condition. The heavy spot is in the center of the tire and this causes wheel hop and tramp when the tire is in motion.

   Answer B is correct. Vibration when braking is a symptom of loose wheel bearings; it would not be caused by improper wheel balance.

   Answer C is incorrect. Wheel hop or tramp is a result of static imbalance. The heavy spot of the tire is influenced by centrifugal force. This influences attempts to move the heavy spot on a tangent line away from the wheel axis. This action will lift the wheel assembly off of the road surface when the heavy spot reaches the top of the rotation producing wheel tramp. When the wheel and tire move downward, the tire strikes the road surface with a pounding action. This repeated pounding produces cupping on the tire tread.

   Answer D is incorrect. Wheel shimmy is a result of dynamic imbalance. The heavy spot is toward the outside edge of the tire. In this case, centrifugal force moves the heavy spot toward the centerline of the tire. When the heavy spot is at the rear of the tire, the tire moves inward. As the heavy spot rotates to the front the tire moves outward. This action causes a shimmy that may be felt as steering wheel oscillation at medium and high speeds.

TASK D.11

2. When checking the end-play measurement on a PreSet hub, the tip of the dial indicator should be placed on the hub cap flange or axle mounting flange, and the magnetic base of the dial indicator should be mounted on the hub's:
   A. Spindle end.
   B. Hub cap.
   C. Fill hole.
   D. Axle housing.

   Answer A is correct. The dial indicator tip should be placed on the hub cap flange or axle mounting flange, and the magnetic base of the dial indicator should be mounted on the hub's spindle end.

   Answer B is incorrect. The hub cap should be removed to access the end of the spindle.

   Answer C is incorrect. The fill hole is located in the hub cap. Since the hub cap must be removed, the fill hole is not available.

   Answer D is incorrect. The axle housing is not accessible. The magnetic base should be mounted on the spindle end.

TASK B.9

3. An improperly adjusted air suspension ride height can cause:
   A. Damaging driveline angles.
   B. A hissing noise during compressor operation.
   C. Back pressure in the supply line.
   D. A leak, causing intermittent operation.

Answer A is correct. An improperly adjusted air suspension height control valve can cause a change in driveline angle, causing transmission output bearing failure. Excessive operating angles due to improper leveling valve adjustment cause premature output shaft bearing failure due to harmful torsional vibrations that result from non-uniformity.

Answer B is incorrect. A hissing noise during compressor operation can be caused by a leaking airline.

Answer C is incorrect. The relief valve on the height control will bleed back pressure in the supply line.

Answer D is incorrect. An improperly adjusted height control valve will not cause leaks leading to intermittent conditions.

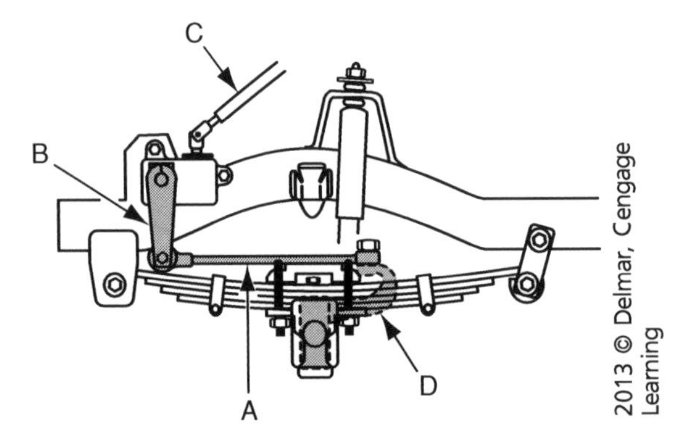

TASK A.18

4. Referring to the figure above, the steering arm is:
   A. Letter A.
   B. Letter B.
   C. Letter C.
   D. Letter D.

Answer A is incorrect. Letter A is the drag link.

Answer B is incorrect. Letter B is the pitman arm.

Answer C is incorrect. Letter C is the steering shaft.

Answer D is correct. Letter D is the steering arm.

5. What is the LEAST LIKELY advantage for the use of sliding fifth wheel assemblies on over-the-road equipment?

TASK B.16

   A. Weight-over-axle on the tractor may be altered.
   B. Weight-over-axle on the trailer may be altered.
   C. The trailer bridge formula may be altered.
   D. It allows for the use of a rigid-mount fifth wheel.

   Answer A is incorrect. The ability to alter weight-over-axle on the tractor is an advantage of sliding fifth wheel assemblies.

   Answer B is incorrect. The ability to alter weight-over-axle on the trailer is an advantage of sliding fifth wheel assemblies.

   Answer C is incorrect. The ability to alter the trailer bridge formula is an advantage of sliding fifth wheel assemblies.

   Answer D is correct. Rigid fifth wheels are not used on over-the-road equipment.

6. All of the following conditions may cause steering wander or pull EXCEPT:

TASK C.1

   A. Improper rear-axle alignment.
   B. Loose steering gear.
   C. Excessive positive caster.
   D. Worn kingpin bushings.

   Answer A is incorrect. If the rear drive axle alignment is out of specification, the axle is positioned to the left or right of the geometric centerline and will cause the vehicle to pull to the left or the right.

   Answer B is incorrect. If the steering gear bolts are loose, it will allow for play in the steering linkage allowing the road forces to direct the vehicle to wander or pull.

   Answer C is correct. Excessive positive caster will not cause the vehicle to wander or pull because the load is projected too far out in front of the vehicle. The tire/wheel assemblies will cause the tires to travel in a straight line and will increase the steering effort when making a turn.

   Answer D is incorrect. Play at the kingpins will allow the front tire/wheel assemblies to move away from a true centerline causing steering instability.

7. When adjusting front-wheel camber:

TASK C.2

   A. If the front wheel has a positive camber angle, the camber line is tilted inward from the tire vertical centerline of the wheel and tire.
   B. Excessive positive camber on a front wheel causes premature wear on the inside edge of the tire tread.
   C. The vehicle will lead to the side that has the most negative camber.
   D. If the camber angle is out of specification, this may be caused by worn suspension components.

   Answer A is incorrect. With positive camber, the camber line is tilted outward from the tire vertical centerline of the wheel and tire.

   Answer B is incorrect. Excessive positive camber places the top of the tire away from the frame. This causes the load to be carried on the outside of the tire tread. The result is tread wear on the outside of the tire tread.

   Answer C is correct. Because of the cone shape effect that occurs with camber angles, the vehicle will always lead to the side with the most positive camber angle.

   Answer D is incorrect. Improper camber angle may be a result of worn suspension components.

TASK D.1

8. Technician A says some abnormal tire wear conditions are normal for certain design tires. Technician B says the TMC Radial Tire Conditions Analysis Guide or Manufacturers Tread Wear Reference Guide should be used to determine tire wear pattern conditions. Who is correct?

   A. A only
   B. B only
   C. Both A and B
   D. Neither A nor B

   Answer A is incorrect. Technician B is also correct.

   Answer B is incorrect. Technician A is also correct.

   Answer C is correct. Both Technicians are correct. Tire application and vehicle vocation will affect tire wear. Vehicles used off-road or in rough terrain will wear the tires differently than those used in highway applications. In those applications, what would be abnormal for highway use would be normal in those applications. The TMC publishes a Radial Tire Conditions Analysis Guide or Manufacturers Tread Wear Reference Guide that will help the technician diagnose the wear pattern and help determine the root cause of the tire issue. It is an excellent resource designed to aid the technician.

   Answer D is incorrect. Both Technicians are correct.

2013 © Delmar, Cengage Learning

TASK B.16

9. Referring to the figure above, the position of a sliding fifth wheel is being discussed. Technician A says that if the fifth wheel is too far ahead, the increased weight transfer to the steer axle might increase steering effort. Technician B says if the fifth wheel is too far rearward, there is not enough weight transfer to the steering axle causing premature front-wheel lockup during a hard brake application. Who is correct?

   A. A only
   B. B only
   C. Both A and B
   D. Neither A nor B

   Answer A is incorrect. Technician B is also correct.

   Answer B is incorrect. Technician A is also correct.

   Answer C is correct. Both Technicians are correct. If the fifth wheel is too far ahead, the steering effort will be increased due to the trailer weight being placed on the frame ahead of the rear suspension midpoint. If the fifth wheel is positioned too far rearward, premature front-wheel lockup may occur during a hard brake action due to trailer weight transfer to the frame behind the rear suspension midpoint.

   Answer D is incorrect. Both Technicians are correct.

TASK B.2

10. How should steering knuckle vertical clearance be adjusted?

    A. Adding or subtracting shims

    B. Selecting the correct draw key

    C. Using the appropriate bushings

    D. Adjusting a stop bolt and jam nut

    Answer A is correct. Adding or subtracting shims increases or decreases the clearance.

    Answer B is incorrect. The draw key secures the kingpin to the axle eye.

    Answer C is incorrect. The bushings have no bearing on the vertical clearance.

    Answer D is incorrect. Adjusting a stop bolt and jam nut have no bearing on vertical clearance.

TASK A.4

11. Of the following scenarios, which one would be the LEAST LIKELY to require a steering sensor recalibration to be performed?

    A. Maintenance or repair work on the steering linkage, steering gear, or other related mechanism

    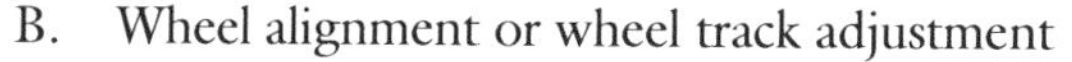

    B. Wheel alignment or wheel track adjustment

    C. Steering wheel replacement

    D. Tire and wheel on vehicle balancing

    Answer A is incorrect. Anytime the steering linkage or steering-related components are repaired or replaced, the steering angle sensor should be recalibrated to ensure the sensor angle and the wheel position are matched.

    Answer B is incorrect. Since alignment and wheel track adjustments may change the position of the tires and system components relative to the steering wheel, a recalibration of the steering sensor should be performed to align the sensor position with the steering linkage.

    Answer C is incorrect. Since the steering wheel is splined to the steering shaft, anytime it is removed or replaced the steering angle sensor should be recalibrated to match the new steering wheel and shaft position.

    Answer D is correct. Tire and wheel balancing will not affect tire and steering component position relative to the steering sensor. Therefore, there is no need to perform a recalibration of the steering angle sensor.

TASK A.11

12. Two technicians are discussing power steering pump damage. Technician A says on disassembly that score marks on the pump drive gear indicate damage. Technician B says that if an overheating condition exists, you must disassemble a power steering pump to determine the extent of any damage. Who is correct?

    A. A only

    B. B only

    C. Both A and B

    D. Neither A nor B

    Answer A is incorrect. Technician B is also correct.

    Answer B is incorrect. Technician A is also correct.

    Answer C is correct. Both Technicians are correct. A damaged power steering pump can be identified by score marks in the drive gear. A power steering pump must be disassembled to observe signs of failure or damage from heat.

    Answer D is incorrect. Both Technicians are correct.

TASK C.3

13. All of the following statements about negative caster are true EXCEPT:

    A. Negative caster does not help return the front wheels to the straight-ahead position after a turn.

    B. Negative caster contributes to directional instability and reduced directional control.

    C. Negative caster contributes to front-wheel shimmy.

    D. Negative caster reduces road shock transmitted to the suspension and chassis.

Answer A is incorrect. Because the load is projected behind the kingpin, the steering effort going into a turn is reduced and the wheels will not return to the straight-ahead position automatically.

Answer B is incorrect. With negative caster, the centerline of the kingpin is behind the vertical centerline of the wheel and spindle at the road surface. If this condition is present, the friction of the tire causes the tire to pivot around the point where the centerline of the kingpin meets the road surface. When this occurs, the wheel is pulled away from the straight-ahead position, which decreases directional stability.

Answer C is correct. Negative caster reduces front-wheel shimmy because the front wheels roll into a road depression without the caster line being aimed at the hole in the road.

Answer D is incorrect. Negative castor reduces road shock because the front wheel is allowed to roll through the irregularity instead of being aimed right at it. This reduces road shock and reduces the amount of shock that is transmitted to the suspension and frame.

TASK B.10

14. A driver complains that after his vehicle is driven on a stretch of rough road, the steering wheel continues to shake for a few seconds. This could be caused by:

    A. Leaking front shock absorbers.

    B. Low tire pressure.

    C. A rusted rear shock absorber.

    D. A missing jounce bumper.

Answer A is correct. Leaking, worn-out front shock absorbers can cause the steering wheel to shake for a few seconds after a stretch of rough road.

Answer B is incorrect. Low tire pressure would not cause wheel shake.

Answer C is incorrect. Rusted rear shock absorbers might contribute to a rough ride due to excessive spring oscillation or excessive wheel hop, but would not cause wheel shake.

Answer D is incorrect. A missing jounce bumper would cause a bang on a rough road but not a shake.

15. Technicians are discussing wheel and tire balance procedures. Technician A says that dynamic wheel balancing is performed with the wheel mounted on the axle. Technician B says that prior to removal from the hub, the position of the dynamically balanced tires should be noted and the tire/wheel assembly should be installed in the same position. Who is correct?

TASK D.7

   A. A only
   B. B only
   C. Both A and B
   D. Neither A nor B

   Answer A is incorrect. Technician B is also correct.

   Answer B is incorrect. Technician A is also correct.

   Answer C is correct. Both Technicians are correct. With the wheel mounted on the axle, a balancer will spin the wheel to determine the location and amount of weight required to balance the tire. A tire that is balanced on the axle should be marked for position before removal because the wheel assembly is balanced but not the tire. The wheel is balanced with all of the rotating components of the wheel end.

   Answer D is incorrect. Both Technicians are correct.

16. Which of the following would cause a featheredged wear pattern across the tire tread?

TASK C.4

   A. Incorrect toe setting
   B. Incorrect camber setting
   C. Incorrect caster setting
   D. Incorrect tire pressure

   Answer A is correct. Improper toe setting causes the tires to scuff while traveling down the road. This scuffing action produced the featheredge wear pattern across the tread face.

   Answer B is incorrect. Incorrect camber settings would wear either the inside or outside of the tire tread.

   Answer C is incorrect. Caster is a non-tire-wearing angle.

   Answer D is incorrect. Overinflation would cause the center section of the tire tread to wear while underinflation would cause tread wear on the outside edges of the tire.

17. What kind of spindle does a conventional design front axle beam hub use?

TASK B.1

   A. Tapered
   B. Straight
   C. Integral
   D. Unitized

   Answer A is correct. The conventional design uses a tapered spindle and accepts conventional wheel ends.

   Answer B is incorrect. Unitized design front axle beams have straight spindles.

   Answer C is incorrect. Integral speaks the design of a spindle assembly, not the type of spindle.

   Answer D is incorrect. Unitized is a hub assembly design that incorporates a straight spindle design.

TASK A.5

18. A driver complains of high turning effort at idle on a vehicle equipped with power steering. Above idle RPM, the system operates normally. What is the LEAST LIKELY cause of this complaint?

A. An unobstructed fluid supply/pressure line
B. Excessive wear between vanes and rotor of the power steering pump assembly
C. A sticking flow control valve
D. An obstruction or kink in the fluid return line

Answer A is correct. An unobstructed power steering pressure line would not cause any adverse power steering conditions, so this is the exception.

Answer B is incorrect. Excessive wear between the vanes and rotor of the power steering pump would cause high turning effort at idle.

Answer C is incorrect. A sticking flow control valve would cause high turning effort at idle.

Answer D is incorrect. An obstruction or kink in the fluid return line would cause high steering effort at idle.

TASK B.3

19. When discussing shock absorbers, all of the statements are true EXCEPT:

A. Shock absorbers control spring action and body sway.
B. Shock absorbers help maintain tire tread contact on the road.
C. Shock absorbers provide more control on the compression cycle than on the extension cycle.
D. Shock absorbers improve vehicle handling and steering control.

Answer A is incorrect. One function of the shock absorber is to control the energy created by the spring movement when the suspension moves up and down. This limits wheel hop and directional instability.

Answer B is incorrect. Because the shock absorber limits the amount of wheel jounce and rebound, the tire tread maintains contact with the road surface improving traction and steerability.

Answer C is correct. Precise control of the oil movement within a shock is the same on the compression cycle and the extension cycle.

Answer D is incorrect. Because of the control of the energy created by the suspension, the vehicle handling and steering control is improved. This is a result of limiting wheel jounce and rebound, and maintaining the tire tread to road contact.

20. Steering wheel shimmy at high speeds may be caused by:

TASK C.1

A. Too much positive caster.
B. Out-of-balance wheel and tire assemblies.
C. Air in the steering system.
D. Low tire pressure.

Answer A is incorrect. Steering wheel oscillation caused by caster would be felt at lower speeds after the wheel assembly hits a bump in the road surface.

Answer B is correct. Dynamic imbalance would cause steering wheel oscillation at high speeds. The heavy spot is toward the outside edge of the tire. In this case centrifugal force moves the heavy spot toward the centerline of the tire. When the heavy spot is at the rear of the tire, the tire moves inward. As the heavy spot rotates to the front, the tire moves outward. This action produces the shimmy.

Answer C is incorrect. Air in the steering system would cause hard steering due to lack of hydraulic force.

Answer D is incorrect. Low tire pressure causes poor tread contact with the road surface resulting in vague or hard steering.

21. Load distribution is being discussed. Technician A says that if the fifth wheel is too far ahead, more weight is transferred to the steer axle increasing steering effort. Technician B says if the fifth wheel is too far rearward, not enough weight will be transferred to the steer axle causing steering instability. Who is correct?

TASK B.12

A. A only
B. B only
C. Both A and B
D. Neither A nor B

Answer A is incorrect. Technician B is also correct.

Answer B is incorrect. Technician A is also correct.

Answer C is correct. Both Technicians are correct. If the fifth wheel is too far ahead, more weight is transferred to the steer axle increasing steering effort. If the fifth wheel is too far back, not enough weight will be transferred to the steer axle causing steering instability because of the decreased amount of weight on the front suspension.

Answer D is incorrect. Both Technicians are correct.

22. The LEAST LIKELY effect of misaligned rear drive axles is:

TASK C.5

A. Steering wheel returning much too fast after cornering.
B. Steering pull.
C. Abnormal tire wear.
D. Erratic steering.

Answer A is correct. Steering wheel return is a result of caster being set positive. Misaligned drive axles would not affect the caster settings on the front steer axle.

Answer B is incorrect. If the rear axle thrust line is left or right of the geometric centerline of the vehicle, the axle position will cause the vehicle to pull to the left or to the right.

Answer C is incorrect. Misaligned drive axles will cause the tires to scrub while traveling down the road. This will cause premature wear across the tire tread.

Answer D is incorrect. The misaligned drive axle will have a tendency to push the vehicle to the left or right. This condition results in reduced steering control and directional instability.

TASK D.2

23. The driver complains of a popping sound and a shake in the steering above 45 mph. What is the most likely cause?

A. Worn shock absorber bushings

B. Out-of-balance tire

C. Incorrect toe-in setting

D. Separated belt in the tire

Answer A is incorrect. Worn shock absorbers would cause the wheel assembly to hop and there may be a slight vibration felt in the steering wheel, but would not cause a popping sound or shake in the steering.

Answer B is incorrect. An out-of-balance tire would cause a shake but no popping sound.

Answer C is incorrect. An incorrect toe-in setting could cause tire wear but not a popping sound or shake in the steering.

Answer D is correct. A separated belt in the tire could cause a popping sound and shake in the steering.

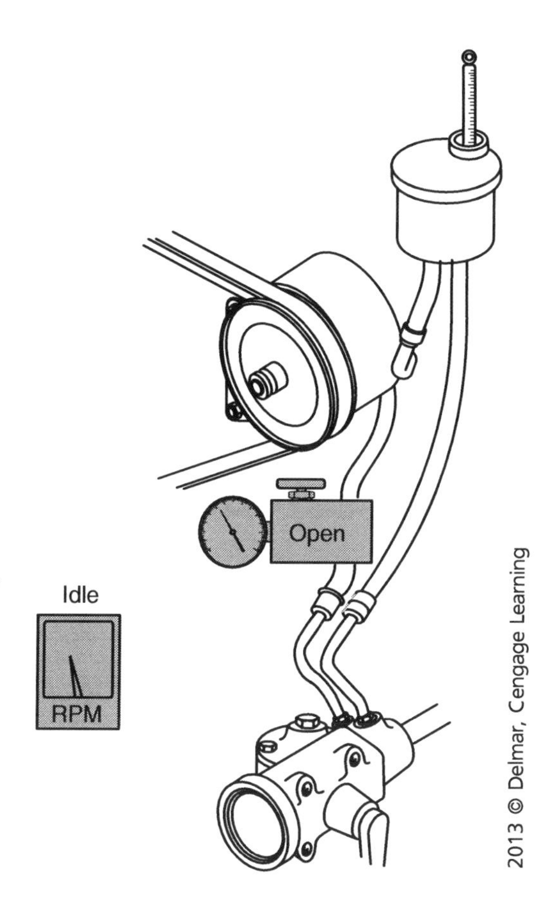

24. Referring to the figure above, two technicians are discussing the procedure for diagnosing a steering system using a steering system analyzer. Technician A states that when the engine is at idle and the gate valve on the analyzer is open, if the pressure is less than 200 psi there may be a restriction in the high-pressure hose. Technician B says that when the flow valve in the analyzer is closed and the pressure is increased to the original equipment manufacturer (OEM) specified test value, if the flow in gallons per minute (GPM) is less than the test value specified by the OEM, the pump may be defective. Who is correct?

TASK A.8

A. A only
B. B only
C. Both A and B
D. Neither A nor B

Answer A is incorrect. Technician B is also correct.

Answer B is incorrect. Technician A is also correct.

Answer C is correct. Both Technicians are correct. Low pressure at idle indicates a restriction before the analyzer in the high-pressure line. If the pump is working properly, both the pressure output and flow rate should be within OEM specifications. When the flow is less this is an indication that the pump may be damaged or worn.

Answer D is incorrect. Both Technicians are correct.

TASK B.3

25. Two technicians are discussing shock absorber inspection. Technician A says that if one shock is defective, both shock absorbers across that axle should be replaced. Technician B says that while performing the inspection, if the shock absorbers are not leaking they are still good. Who is correct?

A. A only

B. B only

C. Both A and B

D. Neither A nor B

Answer A is correct. Only Technician A is correct. Shock absorbers should be replaced in pairs.

Answer B is incorrect. Besides leaking, a shock absorber could be bent, empty, or have a sticking valve.

Answer C is incorrect. Only Technician A is correct.

Answer D is incorrect. Technician A is correct.

TASK A.4

26. A technician is inspecting the steering system for excessive steering wheel free-play on a unit with a 22-inch steering wheel. Which of these statements is correct?

A. Free-play cannot exceed 2 1/4 inches.

B. Free-play cannot exceed 2 3/4 inches.

C. Free-play cannot exceed 3 inches.

D. Free-play cannot exceed 3 1/4 inches.

Answer A is incorrect. Free-play must not exceed 2 3/4 inches.

Answer B is correct. The maximum amount of free-play allowed is 2 3/4 inches.

Answer C is incorrect. Free-play must not exceed 2 3/4 inches. If free-play exceeds 2 3/4 inches inspect the steering components for adjustment, wear, or damage.

Answer D is incorrect. Free-play must not exceed 2 3/4 inches. If free-play exceeds 2 3/4 inches inspect the steering components for adjustment, wear, or damage.

TASK A.19

27. After installing the cross tube assembly, a technician finds that the tie rod clamp interferes with the I-beam on a full turn. If not corrected, which of the following symptoms would result?

A. A noise when turning over bumps

B. Damaged drag link assembly

C. Abnormal steering control

D. Inaccurate camber measurement

Answer A is correct. Tie rod clamp interference with the I-beam on a full turn will cause a noise when turning over bumps.

Answer B is incorrect. A damaged drag link assembly will not result from tie rod clamp interference with the I-beam on a full turn.

Answer C is incorrect. Tie rod clamp interference with the I-beam would not cause abnormal steering control.

Answer D is incorrect. Tie rod clamp interference with the I-beam would not have any effect on front wheel camber.

28. What is the LEAST LIKELY benefit that wheel bearing lubrication provides?

TASK D.10

A. Prevents contamination
B. Cools the bearings
C. Cleans the bearings
D. Prevents corrosion

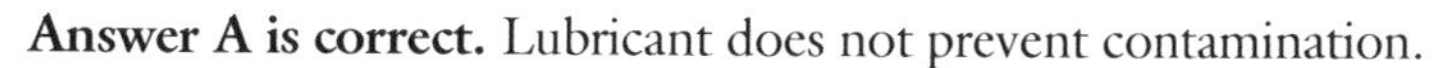

Answer A is correct. Lubricant does not prevent contamination.

Answer B is incorrect. One function of lubricant is to cool the component it is lubricating.

Answer C is incorrect. Lubricants generally have additives that help clean the components.

Answer D is incorrect. Corrosion inhibitors are added to lubricants to help decrease degradation for water or other contaminants.

29. Two technicians are discussing trailer air suspension systems with an external dock lock mechanism. All of the statements are true EXCEPT:

TASK B.7

A. During normal operation, air pressure from the trailer air supply system is supplied to an EDL chamber.
B. When the trailer parking brake is applied, air pressure is released from the EDL chamber.
C. When the flip plates are in the upward position, they are in contact with the air suspension system.
D. The flip plates maintain the trailer at dock height and prevent the trailer from walking forward.

Answer A is incorrect. The EDL chamber receives its air pressure from the trailer air supply system. The air pressure forces the chamber diaphragm to extend the pushrod, moving the flip plates upward.

Answer B is incorrect. When the park brake is set, the air pressure in the EDL chamber is exhausted. The pushrod detracts allowing the flip plates to move downward and engage the air suspension system.

Answer C is correct. When the flip plates are in the upward position, they are disengaged from the suspension system.

Answer D is incorrect. The purpose of the flip plates is to maintain the trailer at dock height, neutralize the air suspension system, and prevent trailer dock walk while the trailer is being loaded.

30. Worn or bent steering column components may cause all of the following EXCEPT:

TASK A.1

A. A rattling noise.
B. Excessive steering free-play.
C. A binding condition when turning the steering wheel.
D. Improper front suspension alignment.

Answer A is incorrect. When steering column components become worn, there is excessive movement within the components. These worn components may be present as a rattling noise, especially over rough road surfaces.

Answer B is incorrect. If the steering column components become worn or bent, the tolerances between the components will increase. This may translate into excessive movement in the steering wheel.

Answer C is incorrect. When components become bent they do not rotate on a true axis. This would allow for friction between the components and would be felt as a binding in the steering column when the steering wheel is rotated.

Answer D is correct. Worn steering column components have no bearing on suspension and alignment specifications or issues.

TASK C.2

31. While discussing camber alignment angles and their effect on a heavy-duty tractor, Technician A says that camber should be set slightly positive. Technician B says that setting camber slightly positive will compensate for normal deflection of the axle and front suspension. Who is correct?

A. A only

B. B only

C. Both A and B

D. Neither A nor B

Answer A is incorrect. Technician B is also correct.

Answer B is incorrect. Technician A is also correct.

Answer C is correct. Both Technicians are correct. Most vehicles with an I-beam front suspension set the camber angle slightly positive. This helps with steerability issues due to varying driving conditions such as road crown. By setting camber slightly positive, when a load is placed on the vehicle the front wheels will assume a zero camber angle.

Answer D is incorrect. Both Technicians are correct.

TASK B.8

32. When discussing lift axles with air suspension and air lift, all of the following statements are true EXCEPT:

A. Two quick-release valves are connected in the air system.

B. The air system incorporates a pressure protection valve.

C. Air pressure to the suspension air bags may be adjusted.

D. When the manual valve or toggle switch is in the on position, air is supplied to the lift air bags to raise the axle.

Answer A is incorrect. Some lift axles with air suspension and air lift incorporate two quick-release valves. One valve controls the air pressure for the suspension air bags and the other controls air pressure to the lift system air bags.

Answer B is incorrect. The lift axle air supply system incorporates a pressure protection valve which will close in the event of an air system failure. This protects the brake air supply system.

Answer C is incorrect. The amount of air supplied to the suspension air bags may be adjusted in accordance with the vehicle load.

Answer D is correct. When the manual valve is in the on position, air is exhausted from the lift air bags and air is supplied to the suspension air bags.

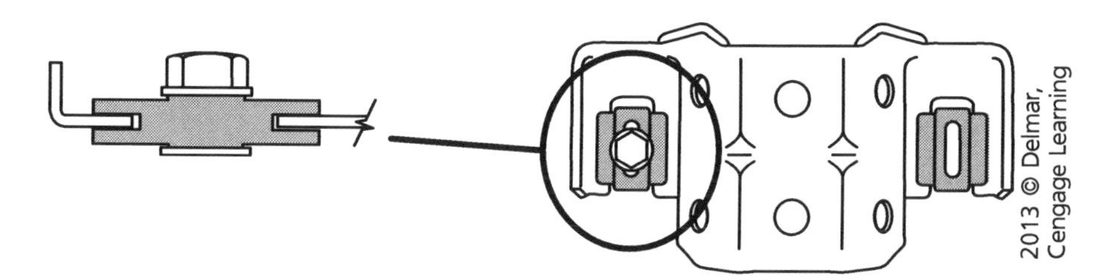

33. Referring to the figure above, two technicians are talking about inspection of collapsible steering columns. Technician A says to check the contact between the bolt head and the bracket. If the bolt head contacts the bracket, the shear load is too high and the column must be replaced. Technician B says to measure the clearance between the capsules and the slots in the steering column bracket. If this measurement is not within specifications, replace the bracket.

A. A only
B. B only
C. Both A and B
D. Neither A nor B

Answer A is incorrect. If the bolt head contacts the bracket, the shear load is too high, but the bracket must be replaced, not the whole column.

Answer B is correct. Only Technician B is correct. If the clearance between the capsules and the slots in the steering column bracket is not within specifications, replace the bracket.

Answer C is incorrect. Only Technician B is correct.

Answer D is incorrect. Technician B is correct.

34. While performing a front-end alignment, a technician notes that the caster angle on the left side of the vehicle is 2 degrees and on the right side of the vehicle the caster angle is 4 degrees. Technician A says that this might indicate that the front axle is twisted. Technician B says that it is possible that the caster shims underneath the springs are of unequal thickness. Who is correct?

A. A only
B. B only
C. Both A and B
D. Neither A nor B

Answer A is incorrect. Technician B is also correct.

Answer B is incorrect. Technician A is also correct.

Answer C is correct. Both Technicians are correct. When a difference is found in the caster angle from left to right, and the caster shims underneath the springs are of equal thickness, this would indicate that the front axle might be twisted. If shims of unequal thickness are installed, this would cause the caster angle to have this condition. Caster shims should be within 1/2 degree of each other.

Answer D is incorrect. Both Technicians are correct.

TASK B.17

35. When the pintle hook is closed, the pintle hook safety latch is held in the lock position:
    A. Electrically.
    B. By air pressure.
    C. By spring tension.
    D. Manually by a lever.

Answer A is incorrect. The pintle hook safety latch is not operated electrically. It maintains its operating position by spring pressure.

Answer B is incorrect. Although air pressure is used to engage the pintle hook ram, it is not used to maintain the pintle hook safety latch.

Answer C is correct. The pintle hook safety latch is maintained in the closed position by spring pressure.

Answer D is incorrect. There is no manual lever associated with the pintle hook assembly.

TASK B.13

36. The LEAST LIKELY cause of a bowed frame rail is:
    A. Operating a dump truck with the box up and loaded.
    B. Snow plow operation.
    C. Vehicle overload.
    D. Unequal loading of the frame.

Answer A is incorrect. Operating a dump truck with the loaded bed in the up position places stresses at the front and rear of the frame fails. These stresses will cause the middle section of the frame rails to bow in an upward direction.

Answer B is incorrect. When the snow plow is pushing a load out in front of the vehicle, the stresses cause the center section of the frame to bow.

Answer C is correct. Overloaded trailers that distribute the weight equally to both frame rails will cause the rails to sag downward.

Answer D is incorrect. An unequal load distribution may cause one frame rail to bow when road stresses are transmitted to the frame.

TASK C.3

37. The LEAST LIKELY affect of positive caster is:
    A. Positive castor reduces steering effort while turning.
    B. Correct positive caster provides improved directional stability of the vehicle.
    C. Excessive positive caster produces harsh riding quality.
    D. When there is a difference between the caster setting for the left and right front tires, the vehicle will pull to the most positive side.

Answer A is correct. Positive caster increases the steering effort while cornering because the driver must overcome the tendency of the tires to remain in a straight-ahead position.

Answer B is incorrect. One of the reasons that caster is set to a positive degree is to improve directional stability. The front tires will automatically maintain a straight-ahead position with positive caster.

Answer C is incorrect. A small amount of positive caster allows the front wheels to roll into a road depression. Excessive positive caster will direct the load right at the hole in the road. This results in road shock being transmitted through the kingpin to the suspension and chassis.

Answer D is incorrect. When there is a difference in the caster angle from left to right, the vehicle will always pull to the most positive side.

38. What connects a pitman arm to the steering control arm?

TASK A.17

A. Drag link
B. Tie rod assembly
C. Steering knuckle
D. Ackerman arm

Answer A is correct. The drag link connects the pitman arm to the steering control arm.

Answer B is incorrect. The tie rod assembly connects the left and right lower steering arms.

Answer C is incorrect. The steering knuckle houses the kingpin assembly.

Answer D is incorrect. The Ackerman arm attaches to the tie rod assembly.

39. When adjusting power steering gears, Technician A states that the wormshaft bearing preload adjustment should be checked with a torque wrench while rotating the gear stub shaft. Technician B states that the pitman shaft over center preload adjustment should be performed with the intermediate steering shaft connected to the steering gear. Who is correct?

TASK A.14

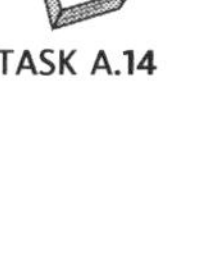

A. A only
B. B only
C. Both A and B
D. Neither A nor B

Answer A is correct. Only Technician A is correct. The wormshaft bearing preload adjustment must be checked with a torque wrench while rotating the gear stub shaft. If the measurement is incorrect, follow the OEM procedures for adjustment.

Answer B is incorrect. The intermediate steering shaft should be disconnected prior to performing the preload adjustment.

Answer C is incorrect. Only Technician A is correct.

Answer D is incorrect. Technician A is correct.

40. Technician A says a bent pitman arm will cause toe-out on turns to be incorrect. Technician B says a bent tie rod will cause an incorrect turning angle. Who is correct?

TASK C.6

A. A only
B. B only
C. Both A and B
D. Neither A nor B

Answer A is incorrect. The bent pitman arm will cause the steering wheel to be off center but will not have any effect on turning radius or toe-out on turns.

Answer B is correct. Only Technician B is correct. A bent tie rod will change the toe angle and will also change the turning radius angle.

Answer C is incorrect. Only Technician B is correct.

Answer D is incorrect. Technician B is correct.

TASK D.10

41. Two technicians are discussing bearing characteristics. All of the statements concerning bearings are true EXCEPT:

A. Cylindrical ball bearings are designed primarily to withstand radial loads, but these bearings may also handle considerable thrust loads.

B. Cylindrical roller bearings are designed primarily to carry radial loads, but they can also handle some thrust load.

C. Tapered roller bearings have excellent radial, thrust, and angular load-carrying capabilities.

D. Needle roller bearings are very compact, and they are designed to carry thrust loads. They will not carry radial loads.

Answer A is incorrect. When a bearing load is applied in a vertical direction, the weight is applied straight down on the bearing. These forces are referred to as radial forces. Forces applied in a horizontal direction are called thrust loads. Because of the ball bearing design, these bearings can handle both stress loads because the load may be transmitted in numerous directions.

Answer B is incorrect. Like cylindrical ball bearings, cylindrical roller bearings are designed to handle high radial thrust. However, thrust load characteristics are reduced due to the rotating design of the bearing.

Answer C is incorrect. Tapered roller bearings have an advantage of handling radial, thrust, and angular loads. This is why they are used in wheel hubs.

Answer D is correct. Needle bearings are designed to handle radial loads and not thrust loads.

TASK B.16

42. A driver complains that the air slide release fifth wheel is hard to release. Technician A says that raising the landing gear to relieve pressure on the plungers should help. Technician B says that if the locking plungers will not release, the problem may be a defective air cylinder. Who is correct?

A. A only

B. B only

C. Both A and B

D. Neither A nor B

Answer A is incorrect. When an air slide release fifth wheel is being released, you lower the landing gear to relieve pressure on the plungers.

Answer B is correct. Only Technician B is correct. If the locking plungers will not release, the air cylinder may be defective.

Answer C is incorrect. Only Technician B is correct.

Answer D is incorrect. Technician B is correct.

43. While performing a front-end alignment, a technician finds that the left-front wheel on an I-beam front suspension has excessive negative camber. This may be caused by:

TASK C.2

A. Improper shim thickness between the underside of the spring and the axle.
B. Worn knuckle pin bushings on the left end of the axle.
C. Seized bearing between the steering knuckle and the axle eye.
D. Worn rear spring shackles and pins.

Answer A is incorrect. Shims placed between the bottom of the spring and the top of the axle are used to adjust caster not camber.

Answer B is correct. Worn knuckle pin bushings will allow the spindle to move inboard toward the frame rail. This would result in a negative camber angle.

Answer C is incorrect. A seized bearing between the steering knuckle and axle eye would cause a noisy operation when cornering and a binding of the steering components. It has no effect on camber angles.

Answer D is incorrect. Worn rear spring shackle pins and bushings might affect caster angles because the frame rail is lowered due to the wear. It will not cause camber to be more negative.

44. Tire matching is being discussed. Technician A says excessive tire wear may be a result of mismatching. Technician B says mismatched tire sizes across a drive axle may cause high axle lubricant temperature. Who is correct?

TASK D.8

A. A only
B. B only
C. Both A and B
D. Neither A nor B

Answer A is incorrect. Technician B is also correct.

Answer B is incorrect. Technician A is also correct.

Answer C is correct. Both Technicians are correct. Mismatched tire sizes can cause excessive tire wear and high axle lubricant temperature due to constant differential gear operation.

Answer D is incorrect. Both Technicians are correct.

45. Erratic torque once every 360 degrees through mesh while rotating the steering wheel is caused by:

TASK A.5

A. A bent worm gear in the steering gear box.
B. Loose wheel lug nuts.
C. Premature tie rod wear.
D. Constant drag when turning the steering wheel.

Answer A is correct. A bent worm gear will cause erratic torque or binding once during every 360 degrees of rotation.

Answer B is incorrect. Loose wheel lug nuts and tie rod wear cannot be caused by a bent worm gear. These are separate operational and regular wear conditions that are not related to the steering gear.

Answer C is incorrect. Premature tie rod wear might cause wander and weave issues while the vehicle is in motion, but not erratic torque issues.

Answer D is incorrect. Continuous drag when turning the steering wheel might be caused by misadjusted steering gear preload or binding steering components.

TASK C.3

46. Two technicians are discussing front-wheel caster angles. Technician A says a vehicle has better directional control when the front axle caster angle is set positive. Technician B says that when the front axle positive caster angle is increased, the ride quality of the vehicle improves. Who is correct?

    A. A only
    B. B only
    C. Both A and B
    D. Neither A nor B

Answer A is correct. Only Technician A is correct. Caster is generally set positive to improve directional control while driving straight ahead and to assist the front wheels to return to a straight-ahead position after cornering.

Answer B is incorrect. Increasing positive caster angles will impair the ride characteristics of the vehicle. Road shock is transmitted through the kingpin to the suspension and chassis. This will cause the front wheels to oscillate at lower speeds producing front-wheel shimmy.

Answer C is incorrect. Only Technician A is correct.

Answer D is incorrect. Technician A is correct.

TASK B.4

47. A vehicle is found to be "dog tracking." Upon inspection, the technician finds the center bolt broken in the spring assembly. Technician A says that worn or loose spring shackles could be the cause. Technician B says that loose U-bolts could be the cause. Who is correct?

    A. A only
    B. B only
    C. Both A and B
    D. Neither A nor B

Answer A is incorrect. Loose spring shackles will not break the spring center bolt.

Answer B is correct. Only Technician B is correct. Loose U-bolts can break the spring center bolt. If the spring U-bolts are not tightened properly, the axle may shift on the spring. This action may shear off the center bolt. This condition causes improper axle position on the spring, resulting in serious steering problems such as pulling to one side and pulling while braking.

Answer C is incorrect. Only Technician B is correct.

Answer D is incorrect. Technician B is correct.

TASK D.3

48. Gold/yellow color wheel studs are used in what hub assembly?

    A. All hub-piloted 10-stud hub assemblies
    B. All hub-piloted eight-stud hub assemblies
    C. All right-hand ball-seat nut hub assemblies
    D. All left-hand ball-seat nut hub assemblies

Answer A is incorrect. Hub-piloted assemblies use a conical washer and nut assembly to retain the wheel to the hub. They are all right-handed thread.

Answer B is incorrect. Gold/yellow wheel studs are used on all left-hand ball-seat nut hub assemblies. Regardless of the number of mounting studs (eight or 10) hub-piloted assemblies use a conical washer and nut assembly to retain the wheel to the hub. They are all right-handed thread.

Answer C is incorrect. Right-hand ball-seat nuts are silver in color.

Answer D is correct. Gold/yellow wheel studs are used on all left-hand ball-seat nut hub assemblies.

49. Which statement is correct concerning turning radius (toe-out on turns)?

TASK C.6

A. Turning radius is adjusted by lengthening or shortening the tie rod.
B. The stop bolts are adjusted to limit it.
C. Turning radius will affect tire tread wear.
D. The inside wheel is parallel to the outside wheel when cornering.

Answer A is incorrect. Adjusting the length of the tie rod affects the toe angle of the front wheels.

Answer B is correct. The axle stop bolts are adjusted to limit the amount of turning radius the front wheel may have.

Answer C is incorrect. If the turning radius is correct, there should be no affect on tire wear.

Answer D is incorrect. The inside wheel turns at a sharper angle than the outside wheel. If they were parallel to each other tire scrub would occur.

50. When performing an axle alignment on a vehicle equipped with an air ride suspension, Technician A states that the frame height should be checked prior to performing the alignment. Technician B says that the rear axle should be aligned to the steer axle. Who is correct?

TASK B.7

A. A only
B. B only
C. Both A and B
D. Neither A nor B

Answer A is correct. Only Technician A is correct. Frame height should be checked prior to axle alignment to ensure proper vehicle attitude.

Answer B is incorrect. The rear axle is aligned to the frame, not to the front axle.

Answer C is incorrect. Only Technician A is correct.

Answer D is incorrect. Technician A is correct.

SECTION

7 Appendices

PREPARATION EXAM ANSWER SHEET FORMS

ANSWER SHEET

1. ____________
2. ____________
3. ____________
4. ____________
5. ____________
6. ____________
7. ____________
8. ____________
9. ____________
10. ____________
11. ____________
12. ____________
13. ____________
14. ____________
15. ____________
16. ____________
17. ____________
18. ____________
19. ____________
20. ____________
21. ____________
22. ____________
23. ____________
24. ____________
25. ____________
26. ____________
27. ____________
28. ____________
29. ____________
30. ____________
31. ____________
32. ____________
33. ____________
34. ____________
35. ____________
36. ____________
37. ____________
38. ____________
39. ____________
40. ____________
41. ____________
42. ____________
43. ____________
44. ____________
45. ____________
46. ____________
47. ____________
48. ____________
49. ____________
50. ____________

ANSWER SHEET

1. ______________
2. ______________
3. ______________
4. ______________
5. ______________
6. ______________
7. ______________
8. ______________
9. ______________
10. ______________
11. ______________
12. ______________
13. ______________
14. ______________
15. ______________
16. ______________
17. ______________
18. ______________
19. ______________
20. ______________
21. ______________
22. ______________
23. ______________
24. ______________
25. ______________
26. ______________
27. ______________
28. ______________
29. ______________
30. ______________
31. ______________
32. ______________
33. ______________
34. ______________
35. ______________
36. ______________
37. ______________
38. ______________
39. ______________
40. ______________
41. ______________
42. ______________
43. ______________
44. ______________
45. ______________
46. ______________
47. ______________
48. ______________
49. ______________
50. ______________

ANSWER SHEET

1. ______________
2. ______________
3. ______________
4. ______________
5. ______________
6. ______________
7. ______________
8. ______________
9. ______________
10. ______________
11. ______________
12. ______________
13. ______________
14. ______________
15. ______________
16. ______________
17. ______________
18. ______________
19. ______________
20. ______________
21. ______________
22. ______________
23. ______________
24. ______________
25. ______________
26. ______________
27. ______________
28. ______________
29. ______________
30. ______________
31. ______________
32. ______________
33. ______________
34. ______________
35. ______________
36. ______________
37. ______________
38. ______________
39. ______________
40. ______________
41. ______________
42. ______________
43. ______________
44. ______________
45. ______________
46. ______________
47. ______________
48. ______________
49. ______________
50. ______________

ANSWER SHEET

1. ______________
2. ______________
3. ______________
4. ______________
5. ______________
6. ______________
7. ______________
8. ______________
9. ______________
10. ______________
11. ______________
12. ______________
13. ______________
14. ______________
15. ______________
16. ______________
17. ______________
18. ______________
19. ______________
20. ______________
21. ______________
22. ______________
23. ______________
24. ______________
25. ______________
26. ______________
27. ______________
28. ______________
29. ______________
30. ______________
31. ______________
32. ______________
33. ______________
34. ______________
35. ______________
36. ______________
37. ______________
38. ______________
39. ______________
40. ______________
41. ______________
42. ______________
43. ______________
44. ______________
45. ______________
46. ______________
47. ______________
48. ______________
49. ______________
50. ______________

ANSWER SHEET

1. ______
2. ______
3. ______
4. ______
5. ______
6. ______
7. ______
8. ______
9. ______
10. ______
11. ______
12. ______
13. ______
14. ______
15. ______
16. ______
17. ______
18. ______
19. ______
20. ______
21. ______
22. ______
23. ______
24. ______
25. ______
26. ______
27. ______
28. ______
29. ______
30. ______
31. ______
32. ______
33. ______
34. ______
35. ______
36. ______
37. ______
38. ______
39. ______
40. ______
41. ______
42. ______
43. ______
44. ______
45. ______
46. ______
47. ______
48. ______
49. ______
50. ______

ANSWER SHEET

1.	______	21.	______	41.	______
2.	______	22.	______	42.	______
3.	______	23.	______	43.	______
4.	______	24.	______	44.	______
5.	______	25.	______	45.	______
6.	______	26.	______	46.	______
7.	______	27.	______	47.	______
8.	______	28.	______	48.	______
9.	______	29.	______	49.	______
10.	______	30.	______	50.	______
11.	______	31.	______		
12.	______	32.	______		
13.	______	33.	______		
14.	______	34.	______		
15.	______	35.	______		
16.	______	36.	______		
17.	______	37.	______		
18.	______	38.	______		
19.	______	39.	______		
20.	______	40.	______		

Glossary

Ackerman Principle The geometric principle used to provide toe-out on turns. The ends of the steering arms are angled so that the inside wheel turns more than the outside wheel when a vehicle is making a turn.

Actuator A device that delivers motion in response to an electrical signal.

Adapter The welds under a spring seat to increase the mounting height or fit a seal to the axle.

Adaptive Suspension A term often used to describe an air spring suspension because the suspension adjusts to load conditions automatically, providing a low-rate suspension with light or no loads and a high-rate suspension with heavier loads. Air ride suspensions may take the place of a leaf spring suspension or be used in conjunction with them.

Additive An ingredient intended to improve a certain characteristic of the material or fluid.

Adjustable Torque Arm A member used to retain axle alignment and, in some cases, control axle torque. Normally one adjustable and one rigid torque arm are used per axle so the axle can be aligned. This rod can be extended or retracted for adjustment purposes.

Air Bag An air-filled device that functions as the spring on axles that utilize air pressure in the suspension system.

Air Compressor (1) An engine-driven mechanism for supplying high-pressure air to the truck brake system. There are basically two types of compressors: those designed to work on in-line engines and those that work on V-type engines. The in-line type is mounted forward and is gear driven, while the V-type is mounted toward the firewall and is camshaft driven. With both types the coolant and lubricant are supplied by the truck engine. (2) A pump-like device in the air conditioning system that compresses refrigerant vapor to achieve a change in state for the refrigeration process.

Air Dryer A unit that removes moisture.

Air Filter/Regulator Assembly A device that minimizes the possibility of moisture-laden air or impurities entering a system.

Air Hose An air line, such as one between the tractor and trailer, that supplies air for the trailer brakes.

Air Spring An air-filled device that functions as the spring on axles that utilize air pressure in the suspension system.

Air Spring Suspension A single- or multi-axle suspension relying on air bags for springs and weight distribution of axles.

Ambient Temperature Temperature of the surrounding or prevailing air. Normally, it is considered to be the temperature in the service area where testing is taking place.

American Trucking Association Practice and standards body that represents American Trucking Industries. Usually known as ATA.

Analog Volt/Ohmmeter (AVOM) A test meter used for checking voltage and resistance. Analog meters should not be used on solid-state circuits.

Anticorrosion A chemical agent used to protect metal surfaces from corrosion.

Applied Moment A term meaning a given load has been placed on a frame at a particular point.

Area The total cross section of a frame rail including all applicable elements usually given in square inches.

Articulation Vertical movement of the front driving or rear axle relative to the frame of the vehicle to which they are attached.

ASE Abbreviation for Automotive Service Excellence, a trademark of the National Institute for Automotive Service Excellence.

Aspect Ratio A tire term calculated by dividing the tire's section height by its section width.

Atmospheric Pressure The weight of the air at sea level; 14.696 pounds per square inch (psi) or 101.33 kilopascals (kPa).

Axis of Rotation The centerline around which a gear or part revolves.

Axle (1) A rod or bar on which wheels turn. (2) A shaft that transmits driving torque to the wheels.

Axle Seat A suspension component used to support and locate the spring on an axle.

Axle Shims Thin wedges that may be installed under the leaf springs of single-axle vehicles to tilt the axle and correct the U-joint operating angles. Wedges are available in a range of sizes to change pinion angles.

Beam Solid Mount Suspension A tandem suspension relying on a pivotal mounted beam, with axles attached at the ends for load equalization. The beam is mounted to a solid center pedestal.

Beam Suspension A tandem suspension relying on a pivotally mounted beam, with axles attached at the ends for load equalization. Beam is mounted to center spring.

Bellows A movable cover or seal that is pleated or folded like an accordion to allow for expansion and contraction.

Bending Moment A term implying that when a load is applied to the frame, it will be distributed across a given section of the frame material.

Bias A tire term where belts and plies are laid diagonally or crisscrossing each other.

Bleed Air Tanks The process of draining condensation from air tanks to increase air capacity and brake efficiency.

Block Diagnosis Chart A troubleshooting chart that lists symptoms, possible causes, and probable remedies in columns.

Bobtailing A tractor running without a trailer.

Bogie The axle spring, suspension arrangement on the rear of a tandem-axle tractor.

Boss A heavy cast section that is used for support, such as the outer race of a bearing or shaft bore.

Bottom U-Bolt Plate A plate that is located on the bottom side of the spring or axle and is held in place when the U-bolts are tightened to the clamp spring and axle together.

Bottoming A condition that occurs when: (1) The teeth of one gear touch the lowest point between teeth of a mating gear. (2) The bed or frame of the vehicle strikes the axle, such as may be the case of overloading.

Bracket An attachment used to secure components to the body or frame.

Bump Steer Erratic steering caused from rolling over bumps, cornering, or heavy braking. Same as orbital steer and roll steer.

Camber The attitude of a wheel and tire assembly when viewed from the front of a car. If it leans outward, away from the car at the top, the wheel is said to have positive camber. If it leans inward, it is said to have negative camber.

Caster The angle formed between the kingpin axis and a vertical axis as viewed from the side of the vehicle. Caster is considered positive when the top of the kingpin axis is behind the vertical axis.

Center of Gravity The point around which the weight of a truck is evenly distributed; the point of balance.

Check Valve A valve that allows air to flow in one direction only. It is a federal requirement to have a check valve between the wet and dry air tanks.

COE Abbreviation for cab-over-engine.

Coefficient of Friction A measurement of the amount of friction developed between two objects in physical contact when one of the objects is drawn across the other.

Coil Springs Spring steel spirals that are mounted on control arms or axles to absorb road shock.

Combination A truck coupled to one or more trailers.

Compression Applying pressure to a spring or any springy substance, thus causing it to reduce its length in the direction of the compressing force.

Compressor (1) A mechanical device that increases pressure within a container by pumping air into it. (2) That component of an air-conditioning system that compresses low-temperature/pressure refrigerant vapor.

Constant Rate Springs Leaf-type spring assemblies that have a constant rate of deflection.

Control Arm The main link between the vehicle's frame and the wheels that acts as a hinge to allow wheel action up and down independent of the chassis.

Converter Dolly An axle, frame, drawbar, and fifth wheel arrangement that converts a semitrailer into a full trailer.

Cross Groove Joint disc-shaped type of inner CV joint that uses balls and V-shaped grooves on the inner and outer races to accommodate the plunging motion of the half-shaft. The joint usually bolts to a transaxle stub flange; same as a disc-type joint.

Cross Tube A system that transfers the steering motion to the opposite, passenger-side steering knuckle. It links the two steering knuckles together and forces them to act in unison.

C-Train A combination of two or more trailers in which the dolly is connected to the trailer by means of two pintle hook or coupler drawbar connections. The resulting connection has one pivot point.

Dampen To slow or reduce oscillations or movement.

Dead Axle Non-live or dead axles are often mounted in lifting suspensions. They hold the axle off the road when the vehicle is traveling empty and put it on the road when a load is being carried. They are also used as air suspension third axles on heavy straight trucks and are used extensively in eastern states with high axle weight laws. It is an axle that does not rotate but merely forms a base on which to attach the wheels.

Deadline To take a vehicle out of service.

Deburring To remove sharp edges from a cut.

Dedicated Contract Carriage Trucking operations set up and run according to a specific shipper's needs. In addition to transportation, they often provide other services such as warehousing and logistics planning.

Deflection Bending or moving to a new position as the result of an external force.

Department of Transportation (DOT) A government agency that establishes vehicle standards.

Dial Caliper A measuring instrument capable of taking inside, outside, depth, and step measurements.

Differential Carrier Assembly An assembly that controls the drive axle operation.

Digital Volt/Ohmmeter (DVOM) A type of test meter recommended by most manufacturers for use on solid-state circuits.

Dispatch Sheet A form used to keep track of dates when the work is to be completed. Some dispatch sheets follow the job through each step of the servicing process.

Dog Tracking Off-center tracking of the rear wheels as related to the front wheels.

DOT Abbreviation for Department of Transportation.

Drag Link A connecting rod or link between the steering gear, pitman arm, and the steering linkage.

Drive Train An assembly that includes all power transmitting components from the rear of the engine to the wheels, including clutch/torque converter, transmission, driveline, and front and rear driving axles.

Driver's Manual A publication that contains information needed by the driver to understand, operate, and care for the vehicle and its components.

Dynamic Balance Dynamic wheel balance distributes the weight of the tire and wheel from side to side across the tire tread.

Electronics The technology of controlling electricity.

Elliot Axle A solid-bar front axle on which the ends span the steering knuckle.

Equalizer A suspension device used to transfer and maintain equal load distribution between two or more axles of a suspension. Formerly called a rocker beam.

Equalizer Bracket A bracket for mounting the equalizer beam of a multiple axle spring suspension to a truck or trailer frame while allowing for the beam's pivotal movement. Normally there are three basic types: flange-mount, straddle-mount, and under- or side-mount.

Extra Capacity A term that generally refers to: (1) A coupling device that has strength capability greater than standard. (2) An oversized tank or reservoir for a fluid or vapor.

False Brinelling The polishing of a surface that is not damaged.

Fatigue Failures The progressive destruction of a shaft or gear teeth material usually caused by overloading.

Federal Motor Vehicle Safety Standard (FMVSS) A federal standard that specifies that all vehicles in the United States be assigned a Vehicle Identification Number (VIN).

Federal Motor Vehicle Safety Standard No. 121 (FMVSS 121) A federal standard that made significant changes in the guidelines that cover air brake systems. Generally speaking, the requirements of FMVSS 121 are such that larger-capacity brakes and heavier steerable axles are needed to meet them.

FHWA Abbreviation for Federal Highway Administration.

Fiber Composite Springs Springs that are made of fiberglass, laminated, and bonded together by tough polyester resins.

Flare To spread gradually outward in a bell shape.

FMVSS Abbreviation for Federal Motor Vehicle Safety Standard.

FMVSS No. 121 Abbreviation for Federal Motor Vehicle Safety Standard No. 121.

Foot-Pound An English unit of measurement for torque. One foot-pound is the torque obtained by a force of 1 pound applied to a foot-long wrench handle.

Frame Width The measurement across the outside of the frame rails of a tractor, truck, or trailer.

Franchised Dealership A dealer that has signed a contract with a particular manufacturer to sell and service a particular line of vehicles.

Fretting A result of vibration that the bearing outer race can pick up the machining pattern.

Front Hanger A bracket for mounting the front of the truck or trailer suspensions to the frame. Made to accommodate the end of the spring on spring suspensions. There are four basic types: flange-mount, straddle-mount, under-mount, and side-mount.

Full Trailer A trailer that does not transfer load to the towing vehicle. It employs a tow bar coupled to a swiveling or steerable running gear assembly at the front of the trailer.

Fully Floating Axles An axle configuration whereby the axle half-shafts transmit only driving torque to the wheels and not bending and torsional loads that are characteristic of the semi-floating axle.

GCW Abbreviation for gross combination weight.

Gross Combination Weight (GCW) The total weight of a fully equipped vehicle including payload, fuel, and driver.

Gross Trailer Weight (GTW) The sum of the weight of an empty trailer and its payload.

Gross Vehicle Weight (GVW) The total weight of a fully equipped vehicle and its payload.

GTW Abbreviation for gross trailer weight.

GVW Abbreviation for gross vehicle weight.

Heavy-Duty Truck A truck that has a GVW of 26,001 pounds or more.

Helper Spring An additional spring device that permits greater load on an axle.

High CG Load Any application in which the load center of gravity (CG) of the trailer exceeds 40 inches (102 cm) above the top of the fifth wheel.

High-Strength Steel A low-alloy steel that is much stronger than hot-rolled or cold-rolled sheet steels that normally are used in the manufacture of car body frames.

I-Beam Axle An axle designed to give great strength at reasonable weight. The cross section of the axle resembles the letter "I."

ICC Check Valve A valve that allows air to flow in one direction only. It is a federal requirement to have a check valve between the wet and dry air tanks.

Inboard Toward the centerline of the vehicle.

Installation Templates Drawings supplied by some vehicle manufacturers to allow the technician to correctly install the accessory. The templates available can be used to check clearances or to ease installation.

Interleaf Friction The friction between the leaves within a suspension spring pack. Interleaf friction provides a self-dampening characteristic to the spring pack.

Jounce The vertical (upward) wheel movement that occurs when the tire and wheel strike a bump in the road surface. The most compressed condition of a spring.

Kinetic Energy Energy in motion.

Kingpin (1) The pin mounted through the center of the trailer upper coupler (bolster plate) that mates with the fifth wheel locks, securing the trailer to the fifth wheel. The configuration is controlled by industry standards. (2) A pin or shaft on which the steering spindle rotates.

Landing Gear The retractable supports for a semitrailer to keep the trailer level when the tractor is detached from it.

Lateral Runout The wobble or side-to-side movement of a rotating wheel or of a rotating wheel and tire assembly.

Laser Beam Alignment System A two- or four-wheel alignment system using wheel-mounted instruments to project a laser beam to measure toe, caster, and camber.

Lead The tendency of a truck to deviate from a straight path on a level road when there is no pressure on the steering wheel in either direction.

Leaf Springs Strips of steel connected to the chassis and axle to isolate the vehicle from road shock.

Less Than Truckload (LTL) Partial loads from the networks of consolidation centers and satellite terminals.

Lift Axle An auxiliary axle that may be lowered and raised to carry additional vehicle weight. They may be a fixed or steerable axle design and may be mechanically or electronically controlled.

Light Beam Alignment System An alignment system using wheel-mounted instruments to project light beams onto charts and scales to measure toe, caster, and camber, and note the results of alignment adjustments.

Linkage A system of rods and levers used to transmit motion or force.

Live Axle An axle on which the wheels are firmly affixed. The axle drives the wheels.

Live Beam Axle A non-independent suspension in which the axle moves with the wheels.

Lockstrap A manual adjustment mechanism that allows for the adjustment of free travel.

Longitudinal Leaf Spring A leaf spring that is mounted so it is parallel to the length of the vehicle.

LTL Abbreviation for less than truckload.

Maintenance Manual A publication containing routine maintenance procedures and intervals for vehicle components and systems.

Moisture Ejector A valve mounted to the bottom or side of the supply and service reservoirs that collects water and expels it every time the air pressure fluctuates.

Mounting Bracket That portion of the fifth wheel assembly that connects the fifth wheel top plate to the tractor frame or fifth wheel mounting system.

Multiaxle Suspension A suspension consisting of more than three axles.

NATEF Abbreviation for National Automotive Technicians Education Foundation.

National Automotive Technicians Education Foundation (NATEF) A foundation having a program of certifying secondary and postsecondary automotive and heavy-duty truck training programs.

National Institute for Automotive Service Excellence (ASE) A nonprofit organization that has an established certification program for automotive, heavy-duty truck, auto body repair, engine machine shop technicians, and parts specialists.

Needlenose Pliers This tool has long tapered jaws for grasping small parts or for reaching into tight spots. Many needlenose pliers also have cutting edges and a wire stripper.

NIASE Abbreviation for National Institute for Automotive Service Excellence, now abbreviated ASE.

NIOSH Abbreviation for National Institute for Occupation Safety and Health.

NHTSA Abbreviation for National Highway Traffic Safety Administration.

NLGI Abbreviation for National Lubricating Grease Institute.

Non-Live Axle Non-live or dead axles are often mounted in lifting suspensions. They hold the axle off the road when the vehicle is traveling empty and put it on the road when a load is being carried. They are also used as air suspension third axles on heavy straight trucks and are used extensively in eastern states with high axle weight laws.

Nose The front of a semitrailer.

OEM Abbreviation for original equipment manufacturer.

Off-road With reference to unpaved, rough, or ungraded terrain on which a vehicle will operate. Any terrain not considered part of the highway system falls into this category.

On-road With reference to paved or smooth-graded surface terrain on which a vehicle will operate, generally considered to be part of the public highway system.

Oscillation The rotational movement in either fore/aft or side-to-side direction about a pivot point. Generally refers to fifth wheel designs in which fore/aft and side-to-side articulation are provided.

OSHA Abbreviation for Occupational Safety and Health Administration.

Out-of-Round A wheel or tire defect in which the wheel or tire is not round.

Oval A condition that occurs when a tube is not round, but is somewhat egg-shaped.

Parts Requisition A form that is used to order new parts on which the technician writes the names of what part(s) are needed along with the vehicle's VIN or company's identification folder.

Payload The weight of the cargo carried by a truck, not including the weight of the body.

Pipe or Angle Extrusions between opposite hangers on a spring or air-type suspension.

Pitman Arm A steering linkage component that connects the steering gear to the linkage at the left end of the center link.

Pitting Surface irregularities resulting from corrosion.

Plies The layers of rubber-impregnated fabric that make up the body of a tire.

Pounds per Square Inch (psi) A unit of English measure for pressure.

Power A measure of work being done.

Power Steering A steering system utilizing hydraulic pressure to reduce the turning effort required of the operator.

PreSet® Hub Design PreSet hub assemblies utilize pre-adjusted bearings to simplify installation, improve seal and bearing life, and reduce maintenance requirements. PreSet hub assemblies use pre-adjusted bearing technology. A more precise bearing setting is achieved by carefully controlling all critical tolerances in the hub, bearings, and the precision spacer.

Pressure The amount of force applied to a definite area measured in pounds per square inch (psi) English or kilopascals (kPa) metric.

Pressure Differential The difference in pressure between any two points of a system or a component.

Pressure Relief Valve (1) A valve located on the wet tank, usually preset at 150 psi (1034 kPa). Limits system pressure if the compressor or governor unloader valve malfunctions. (2) A valve located on the rear head of an air-conditioning compressor or pressure vessel that opens if an excessive system pressure is exceeded.

psi Abbreviation for pounds per square inch.

Pull Circuit A circuit that brings the cab from a fully tilted position up and over the center.

Push Circuit A circuit that raises the cab from the lowered position to the desired tilt position.

Quick-Release Valve A device used to exhaust air as close as possible to the service chambers or spring brakes.

Radial A tire design having cord materials running in a direction from the center point of the tire, usually from bead to bead.

Radial Load A load that is applied at 90 degrees to an axis of rotation.

Rated Capacity The maximum recommended safe load that can be sustained by a component or an assembly without permanent damage.

Rear Hanger A bracket for mounting the rear of a truck or trailer suspension to the frame. Made to accommodate the end of the spring on spring suspensions. There are usually four types: flange-mount, straddle-mount, under-mount, and side-mount.

Rebound Rebound is the downward movement of the tire and wheel assembly after wheel jounce has occurred.

Recall Bulletin A bulletin that pertains to special situations that involve service work or replacement of components in connection with a recall notice.

Resisting Bending Moment A measurement of frame rail strength derived by multiplying the section modulus of the rail by the yield strength of the material. This term is universally used in evaluating frame rail strength.

Reverse Elliot Axle A solid-beam front axle on which the steering knuckles span the axle ends.

Revolutions per Minute (rpm) The number of complete turns a member makes in one minute.

Right to Know Law A law passed by the federal government and administered by the Occupational Safety and Health Administration (OSHA) that requires any company that uses or produces hazardous chemicals or substances to inform its employees, customers, and vendors of any potential hazards that may exist in the workplace as a result of using the products.

Rigid Torque Arm A member used to retain axle alignment and, in some cases, to control axle torque. Normally, one adjustable and one rigid arm are used per axle so the axle can be aligned.

Rocker Beam A suspension device used to transfer and maintain equal load distribution between two or more axles of a suspension.

Roll Axis The theoretical line that joins the roll center of the front and rear axles.

Rotation A term used to describe the fact that a gear, shaft, or other device is turning.

rpm Abbreviation for revolutions per minute.

Runout A deviation of the specified normal travel of an object. The amount of deviation or wobble a shaft or wheel has as it rotates. Runout is measured with a dial indicator.

Safety Factor (SF) (1) The amount of load that can safely be absorbed by and through the vehicle chassis frame members. (2) The difference between the stated and rated limits of a product, such as a grinding disk.

Screw Pitch Gauge A gauge used to provide a quick and accurate method of checking the threads per inch of a nut or bolt.

Section Height The tread center to bead plane on a tire.

Section Width The measurement on a tire from sidewall to sidewall.

Semifloating Axle An axle type whereby drive power from the differential is taken by each axle half-shaft and transferred directly to the wheels. A single bearing assembly, located at the outer end of the axle, is used to support the axle half-shaft.

Semitrailer A load-carrying vehicle equipped with one or more axles and constructed so that its front end is supported on the fifth wheel of the truck or tractor that pulls it.

Sensor An electronic device used to monitor relative conditions for computer control requirements.

Service Bulletin A publication that provides the latest service tips, field repairs, product improvements, and related information of benefit to service personnel.

Service Manual A manual, published by the manufacturer, that contains service and repair information for all vehicle systems and components.

Shock Absorber A hydraulic device used to dampen vehicle spring oscillations for controlling body sway and wheel bounce, and/or prevent spring breakage.

Single-Axle Suspension A suspension with one axle.

Solvent A substance that dissolves other substances.

Spalling Surface fatigue that occurs when chips, scales, or flakes of metal break off due to fatigue rather than wear. Spalling is usually found on splines and U-joint bearings.

Specialty Service Shop A shop that specializes in areas such as engine rebuilding, transmission/axle overhauling, brake, air conditioning/heating repairs, and electrical/electronic work.

Splined Yoke A yoke that allows the drive shaft to increase in length to accommodate movements of the drive axles.

Spread Tandem Suspension A two-axle assembly in which the axles are spaced to allow maximum axle loads under existing regulations. The distance is usually more than 55 inches.

Spring A device used to reduce road shocks and transfer loads through suspension components to the frame of the trailer. There are usually four basic types: multileaf, monoleaf, taper, and air springs.

Spring Chair A suspension component used to support and locate the spring on an axle.

Spring Deflection The depression of a trailer suspension when the springs are placed under load.

Spring Rate The load required to deflect the spring a given distance (usually 1 inch).

Spring Spacer A riser block often used on top of the spring seat to obtain increased mounting height.

Stability A relative measure of the handling characteristics that provide the desired and safe operation of the vehicle during various maneuvers.

Stabilizer A device used to stabilize a vehicle during turns; sometimes referred to as a sway bar.

Stabilizer Bar A bar that connects the two sides of a suspension so that cornering forces on one wheel are shaped by the other. This helps equalize wheel-side loading and reduces the tendency of the vehicle body to roll outward in a turn.

Static Balance Balance at rest, or still balance. It is the equal distribution of the weight of the wheel and tire around the axis of rotation so that the wheel assembly has no tendency to rotate by itself regardless of its position.

Steering Angle Sensor The steering angle sensor is used in conjunction with electronic steer and braking systems. It is located in the steering column and monitors the degree of steering wheel movement off of center.

Steering Gear A gear set mounted in a housing that is fastened to the lower end of the steering column used to multiply driver turning force and change rotary motion into longitudinal motion.

Steering Stabilizer A shock absorber attached to the steering components to cushion road shock in the steering system, improving driver control in rough terrain and protecting the system.

Still Balance Balance at rest; the equal distribution of the weight of the wheel and tire around the axis of rotation so that the wheel assembly has no tendency to rotate by itself regardless of its position.

Structural Member A primary load-bearing portion of the body structure that affects its over-the-road performance or crash-worthiness.

Suspension A system whereby the axle or axles of a unit are attached to the vehicle frame, designed in such a manner that road shocks from the axles are dampened through springs reducing the forces entering the frame.

Suspension Height The distance from a specified point on a vehicle to the road surface when not at curb weight.

Sway Bar A component that connects the two sides of a suspension so that cornering forces on one wheel are shared by the other. This helps equalize wheel-side loading and reduces the tendency of the vehicle body to roll outward in a turn.

Tag Axle The rearmost axle of a tandem-axle tractor used to increase the load-carrying capacity of the vehicle.

Tandem One directly in front of the other and working together.

Tandem-Axle Suspension A suspension system consisting of two axles with a means for equalizing weight between them.

Tandem Drive A two-axle drive combination.

Tandem Drive Axle A type of axle that combines two single-axle assemblies through the use of an interaxle differential or power divider and a short shaft that connects the two axles together.

Tie Rod Assembly A system that transfers the steering motion to the opposite, passenger-side steering knuckle. It links the two steering knuckles together and forces them to act in unison.

Time Guide Prepared reference material used for computing compensation payable by the truck manufacturer for repairs or service work to vehicles under warranty or for other special conditions authorized by the company.

TMC The Maintenance Council; division of the ATA that sets service and maintenance standards in the trucking industry.

Toe A suspension dimension that reflects the difference in the distance between the extreme front and rear of the tire.

Toe-In A suspension dimension whereby the front of the tire points inward toward the vehicle.

Toe-Out A suspension dimension whereby the front of the tire points outward from the vehicle.

Toe-Out On Turns Toe-out on turns is the turning angle of the wheel on the inside of the turn compared with the turning angle of the wheel on the outside of the turn.

Top U-Bolt Plate A plate located on the top of the spring and held in place when the U-bolts are tightened to clamp the spring and axle together.

Torque To tighten a fastener to a specific degree of tightness, generally in a given order or pattern if multiple fasteners are involved on a single component.

Torque and Twist A term that generally refers to the forces developed in the trailer and/or tractor frame that are transmitted through the fifth wheel when a rigid trailer, such as a tanker, is required to negotiate bumps like street curbs.

Torque Rod Shim A thin wedge-like insert that rotates the axle pinion to change the U-joint operating angle.

Torsion Bar Suspension A type of suspension system that utilizes torsion bars in lieu of steel leaf springs or coil springs. The typical torsion bar suspension consists of a torsion bar, front crank, and rear crank with associated brackets, a shackle pin, and assorted bushings and seals.

Torsional Rigidity A component's ability to remain rigid when subjected to twisting forces.

Tracking The directional travel of the rear wheels in relation to the front wheels.

Tractor A motor vehicle, without a cargo body, that has a fifth wheel and is used for pulling a semitrailer.

Tractor/Trailer Lift Suspension A single-axle air ride suspension with lift capabilities commonly used with steerable axles for pusher and tag applications.

Trailer A platform or container on wheels pulled by a car, truck, or tractor.

Trailer Slider A movable trailer suspension frame that is capable of changing trailer wheelbase by sliding and locking into different positions.

Tree Diagnosis Chart A chart used to provide a logical sequence for what should be inspected or tested when troubleshooting a repair problem.

Triaxle Suspension A suspension consisting of three axles with a means of equalizing weight between axles.

TTMA Abbreviation for Truck and Trailer Manufacturers Association.

Turning Radius Turning angle or radius is the degree of movement from a straight-ahead position of the front wheels to either an extreme right or left position. It is controlled by the Ackerman arms and is established by the OEM according to the vehicle wheel base. It is not adjustable.

TVW Abbreviation for (1) Total vehicle weight. (2) Towed vehicle weight.

U-Bolt A fastener used to clamp the top U-bolt plate, spring, axle, and bottom U-bolt plate together. Inverted (nuts down) U-bolts cross springs when in place; conventional (nuts up) U-bolts wrap around the axle.

Underslung Suspension A suspension in which the spring is positioned under the axle.

Unitized Wheel End The unitized hub is a permanently sealed and lubricated assembly designed to help reduce wheel-end maintenance. Unitized wheel end bearings are lubricated for the life of the hub, bearing, and seal assembly.

Universal Joint (U-joint) A component that allows torque to be transmitted to components that are operating at different angles.

Unsprung Weight The weight of any chassis components not supported by the suspension.

Vacuum Air below atmospheric pressure. There are three types of vacuums important to engine and component function: manifold vacuum, ported vacuum, and venturi vacuum. The strength of these vacuums depends on throttle opening, engine speed, and load.

Validity List A list supplied by the manufacturer of valid bulletins.

Variable Rate Springs Variable rate springs are leaf-type spring assemblies with a variable deflection rate obtained by varying the effective length of the spring assembly.

Vehicle Body Clearance (VBC) The distance from the inside of the inner tire to the spring or other body structures.

VIN Abbreviation for Vehicle Identification Number.

Wet Tank A supply reservoir.

Wheel Alignment The mechanics of keeping all the components of the steering system in the specified relation to each other.

Wheel Balance The equal distribution of weight in a wheel with the tire mounted. It is an important factor that affects tire wear and vehicle control.

Work (1) Forcing a current through a resistance. (2) The product of a force.

Yaw Rate/Lateral Acceleration Sensor A sensor used in conjunction with the steering angle sensor in electronic steering and roll stability systems. This sensor monitors frame articulation and frame angle, and vehicle acceleration.

Yield Strength The highest stress a material can stand without permanent deformation or damage, expressed in pounds per square inch (psi).

Made in the USA
San Bernardino, CA
02 March 2016